P9-BZL-228

EXPLOSIVES

Devices of Controlled Destruction

These and other books are included in the Encyclopedia of Discovery and Invention series:

Airplanes
Anesthetics
Animation
Atoms
Automobiles
Clocks
Computers
Dinosaurs
Explosives
Genetics
Germs
Gravity
Guns
Human Origins
Lasers
Maps
Microscopes
Movies
Phonograph
Photography
Plastics
Plate Tectonics
Printing Press
Radar
Radios
Railroads
Ships
Submarines
Telephones
Telescopes
Television
Vaccines

EXPLOSIVES
Devices of Controlled Destruction

by SEAN M. GRADY

The ENCYCLOPEDIA of
D•I•S•C•O•V•E•R•Y
and **INVENTION**

P.O. Box 289011 SAN DIEGO, CA 92198-9011

Printed in the U.S.A.

Library of Congress Cataloging-in-Publication Data

Grady, Sean M., 1965-
Explosives: devices of controlled destruction / by Sean M. Grady.

p. cm.—(The Encyclopedia of discovery and invention)
Includes bibliographical references and index.
ISBN 1-56006-250-9
1. Explosives—Juvenile literature. [1. Explosives.] I. Title. II. Series.
TP270.5.G73 1995
662'.2—dc20 94-16834
CIP
AC

Contents

Foreword

The belief in progress has been one of the dominant forces in Western Civilization from the Scientific Revolution of the seventeenth century to the present. Embodied in the idea of progress is the conviction that each generation will be better off than the one that preceded it. Eventually, all peoples will benefit from and share in this better world. R.R. Palmer, in his *History of the Modern World*, calls this belief in progress "a kind of nonreligious faith that the conditions of human life" will continually improve as time goes on.

For over a thousand years prior to the seventeenth century, science had progressed little. Inquiry was largely discouraged, and experimentation, almost nonexistent. As a result, science became regressive and discovery was ignored. Benjamin Farrington, a historian of science, characterized it this way: "Science had failed to become a real force in the life of society. Instead there had arisen a conception of science as a cycle of liberal studies for a privileged minority. Science ceased to be a means of transforming the conditions of life." In short, had this intellectual climate continued, humanity's future would have been little more than a clone of its past.

Fortunately, these circumstances were not destined to last. By the seventeenth and eighteenth centuries, Western society was undergoing radical and favorable changes. And the changes that occurred gave rise to the notion that progress was a real force urging civilization forward. Surpluses of consumer goods were replacing substandard living conditions in most of Western Europe. Rigid class systems were giving way to social mobility. In nations like France and the United States, the lofty principles of democracy and popular sovereignty were being painted in broad, gilded strokes over the fading canvasses of monarchy and despotism.

But more significant than these social, economic, and political changes, the new age witnessed a rebirth of science. Centuries of scientific stagnation began crumbling before a spirit of scientific inquiry that spawned undreamed of technological advances. And it was the discoveries and inventions of scores of men and women that fueled these new technologies, dramatically increasing the ability of humankind to control nature—and, many believed, eventually to guide it.

It is a truism of science and technology that the results derived from observation and experimentation are not finalities. They are part of a process. Each discovery is but one piece in a continuum bridging past and present and heralding an extraordinary future. The heroic age of the Scientific Revolution was simply a start. It laid a foundation upon which succeeding generations of imaginative thinkers could build. It kindled the belief that progress is possible as long as there were gifted men and women who would respond to society's needs. When Antonie van Leeuwenhoek observed *Animalcules* (little animals) through his high-powered microscope in 1683, the discovery did not end there. Others followed who would call these "little animals" bacteria and, in time, recognize their role in the process of health and disease. Robert Koch, a German bacteriologist and winner of the Nobel Prize in Physiology and Medicine, was one of these men. Koch firmly established that bacteria are responsible for causing infectious diseases. He identified, among others, the causative organisms of anthrax and tuberculosis. Alexander Fleming, another Nobel Laureate, progressed still further in the quest to understand and control bacteria. In 1928, Fleming discovered penicillin, the antibiotic wonder drug. Penicillin, and the generations of antibiotics that succeeded it, have done more to prevent premature death than any other discovery in the history of humankind. And as civilization hastens toward the twenty-first century, most agree that the conquest of van Leeuwenhoek's "little animals" will continue.

The *Encyclopedia of Discovery and Invention* examines those discoveries and inventions that have had a sweeping impact on life and thought in the modern world. Each book explores the ideas that led to the invention or discovery, and, more importantly, how the world changed and continues to change because of it. The series also highlights the people behind the achievements—the unique men and women whose singular genius and rich imagination have altered the lives of everyone. Enhanced by photographs and clearly explained technical drawings, these books are comprehensive examinations of the building blocks of human progress.

EXPLOSIVES

Devices of Controlled Destruction

EXPLOSIVES

Introduction

In 1851, a Swedish chemist and industrialist named Alfred Nobel began selling a new product he believed would revolutionize mining. Nobel had been seeking a way to stabilize nitroglycerin, a liquid explosive that was notorious for blowing up when jostled. He discovered that by mixing the oily liquid with an absorbent soil called kieselguhr, he could keep the nitroglycerin from randomly exploding. He called the mixture dynamite. Less than a pound of this new explosive could break through as much rock in one second as a team of workers could dig through in a week.

The invention of dynamite ushered in the era of high explosives—chemicals or mixtures of chemicals that burn almost instantaneously, giving off great amounts of heat and gas. As Nobel predicted, dynamite led the way to an unprecedented growth in mining and created many new opportunities in civil engineering. Roadways and railroads were built that ran straight from city to city without having to weave around hills and mountains. And great canals were blasted in the earth to connect

TIMELINE: EXPLOSIVES

1 2 3 4 5 6 7 8 9 10 11

1 ■ ca. A.D. 650
Byzantine alchemists invent Greek fire.

2 ■ ca. 1000
Invention of gunpowder, most probably in China.

3 ■ 1627
German miner Kaspar Weindel uses gunpowder in excavation of royal mines in Hungary.

4 ■ 1805
English chemist John Dalton proposes atomic theory of matter.

5 ■ 1831
English leather merchant William Bickford invents "safety fuse" for lighting gunpowder charges.

6 ■ 1846
Italian chemist Ascanio Sobrero invents nitroglycerin, declares it too dangerous for humanity to use.

7 ■ 1861
Swedish chemist/industrialist Alfred Nobel patents gunpowder primer for nitroglycerin, sells it as "Nobel's Igniter."

8 ■ 1863
German chemist J. Wilbrand develops trinitrotoluene (TNT).

9 ■ 1867
Alfred Nobel invents dynamite by mixing nitroglycerin with kieselguhr; Swedish chemists Ohlsson and Norrbein patent ammonium-nitrate-based explosive.

10 ■ 1871
German chemist Hermann Sprengel develops explosives that can be mixed at a blast site.

11 ■ 1899-1904
Development of PETN and RDX, later used in warfare; TNT becomes inexpensive enough for commercial and military use.

lakes, rivers, and even oceans separated by thousands of square miles of land.

Although explosives proved to be useful tools for the development of modern civilization, they also showed themselves to be devastating tools of war. High explosives like dynamite offered the world's armies a way to increase their military might without increasing the number of soldiers on the battlefield. Eventually, new and more powerful explosives were developed as nations searched for ways to heighten their destructive power.

These days, explosives are used for more tasks than simply breaking apart rocks or destroying enemy targets. Explosives can fuse metal together for ships and railroad tracks, clean dirt out of oil wells, and even pop open automobile air bags in head-on collisions. Because explosives are such an important part of modern civilization, scientists are working on ways to make them safer to handle and use. And others are working on ways to use explosives to create new kinds of industrial-grade diamonds and superconducting ceramics, materials that will be needed as the world's industries enter the twenty-first century.

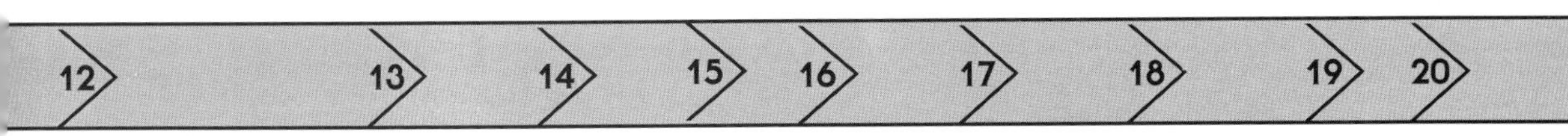

12 ▪ 1914
Outbreak of World War I; first widespread use of TNT as a weapon of war.

13 ▪ ca. 1940
Development of first plastic explosives.

14 ▪ 1947
Cargo ships in Texas City, Texas, and Brest, France, destroyed when their cargoes of ammonium nitrate fertilizer explode; these explosions spark research into development of modern ammonium nitrate explosives.

15 ▪ 1955
Development of Akremite, first practical commercial explosive using ammonium nitrate.

16 ▪ 1957
E. I. du Pont de Nemours begins marketing Tovex-brand water gel explosives.

17 ▪ 1974
Du Pont announces it will stop making dynamite in 1976 to devote more production equipment to making Tovex.

18 ▪ Mid-1970s
Czechoslovakia begins making Semtex, a powerful plastic explosive designed for terrorist use.

19 ▪ 1988
Terrorists plant a plastic explosive bomb on board Pan Am Flight 103; bomb explodes while plane is flying over Lockerbie, Scotland, killing all passengers and crew.

20 ▪ 1990s
Experiments underway to create synthetic industrial diamonds using explosive charges.

CHAPTER 1

Mixing the Black Powder

An explosive is a chemical mixture, usually a solid or a liquid, that burns very fast and very hot, and gives off a great deal of gas. Although the shock of a chemical explosion is a powerful force in its own right, it is the pressure of the gas created by the explosion that does the greatest amount of work or damage. The most powerful explosives can create, in a small area, pressures of more than one million times that of earth's atmosphere, and temperatures equal to that of the sun's surface.

The idea of people controlling forces as powerful as these is almost unbelievable today. A few thousand years ago, the concept of such control was more than unbelievable—it was unknowable. Only natural events such as an erupting volcano or a flashing lightening bolt possessed forces great enough to move rock or shatter a tree. Even so, people around the world sought ways to harness forces such as these for their own benefit.

Taming Fire

These first attempts, which would eventually lead to the invention of explosive devices of controlled destruction, usually involved using fire as a weapon. As far back as 2000 B.C., nations around the Mediterranean Sea and in the Middle East used fire arrows and balls of burning tar to vanquish enemy forces and overrun opposing cities. But these weapons—called incendiary weapons because they were designed to incinerate, or burn, their targets—were not that great a threat. Unless massive volleys were fired, or the flames hit unusually flammable targets like grain or cloth warehouses, incendiaries had only a limited effect. In small numbers, the fires could be put out with little trouble.

In the seventh century, however, the Byzantine Empire introduced the world to a new and devastating form of fire weapon. After experimenting with various compounds, Byzantine alchemists developed a mixture of evergreen resin, sulfur, and crude oil that caught fire quickly and burned fiercely. Once the fiery compound hit a wooden building or ship's deck, it started blazes that were virtually impossible to put out. The Byzantines first used this compound, which they dubbed "Greek fire," to fight off an Arab army that tried to overthrow the Byzantine Empire's capital city, Constantinople.

Greek fire was as great a military revolution in its time as the machine gun was at the turn of the twentieth century. With Greek fire, a small army or group of defenders could hold off a larger force for as long as the supply of the mixture held out. The Byzantine Empire managed to keep the formula for Greek fire a secret for a couple of hundred years. But the formula for a weapon as potent as Greek fire could not be kept secret forever, not so long as the rulers of other nations were will-

The Byzantine army unleashes its powerful new weapon—Greek fire—during the siege of Constantinople.

ing to pay for it. By A.D. 1000, Greek fire had become a common weapon throughout the Western world. It was used extensively in the great wars of the next few centuries, as well as in the Crusades, a series of military campaigns to seize the Holy Lands of the western Mediterranean for Europe.

Gunpowder

Yet for all its devastating effects, Greek fire had a number of serious limitations. Foremost among these limits was firing distance. Greek fire had to be sprayed over the enemy, so the enemy had to be near the soldiers or ships using the weapon. This could be an especially great hazard if the enemy carried its own supply of Greek fire. Also, the many ingredients of Greek fire were not easy to come by or to store. Crude oil, for example, only came from a few areas of the known world, and the nations that controlled the supply were not willing to sell it to their enemies. Gathering pine resin, too, was a time-consuming, arduous task, and transporting it was a matter of braving both robbers and poor roads. Clearly, there was a need for a weapon that had the destructive power of fire yet could be made out of a few easily carried, readily available materials. That weapon was gunpowder.

Gunpowder is a fast-burning mixture of crushed charcoal, saltpeter (also known as potassium nitrate), and sulfur. Its dark black color, which it gets from the large amount of charcoal in the mix, led people to also call it "black powder." No one knows for sure where gunpowder really came from. Its origin has been traced by various historians to

Even with the innovation of machines built especially for hurling Greek fire, the close firing distance made its use hazardous.

A seventeenth-century engraving depicts the production of gunpowder, a fast-burning mixture of crushed charcoal, saltpeter, and sulfur. Its invention spurred an explosives revolution.

China, to the Arab world, and to many of the nations of Europe.

Early Gunpowder Weapons

Although its true origin may never be known, the results of its creation are indisputable. By A.D. 1200, nations from China to England were using gunpowder in war. The earliest gunpowder weapons were little more than glorified fireworks—rockets designed to start fires behind enemy lines and city walls, or short, thick canisters of loosely packed powder that startled enemy soldiers with a quick, thunderous bang. But the armies of the world soon developed weapons that could devastate entire companies of soldiers. One Chinese weapon, with the descriptive title of "leopard-herd-rushing-transversely," fired a volley of twenty-four to thirty-four rocket-propelled arrows from a conical box. When launched, these arrows could travel the length of four football fields and drive through the armor of any foe in their path.

In Europe also, many people discovered gunpowder's ability to propel massive objects. Experimenters found that the force of a gunpowder explosion could be amplified by setting it off inside a long tube, as long as they made the tube strong enough not to be destroyed by the blast. They realized that by forcing the gases in one direction, they could launch heavy objects—for example, a large iron ball—from the gunpowder-filled tubes. This led to the construction of the first mortars—thick-walled buckets of stone or metal that fired heavy boulders or metal balls—and cannons, which at first were merely long-necked mortars with a longer range. The major drawback to these early artillery pieces was their weight. Some of the larger cannons weighed from three to five tons, and could only be drawn by huge teams of oxen. More than once, a retreating army was forced to abandon its artillery to its opponents.

Many soldiers probably would have liked to abandon their cannons even if their side was winning. Serving in or near a battery of cannons was hazardous work. There was always the risk that the cannons themselves could explode if they were loaded with too much powder. There was an even greater risk of an explosion from the gunpowder stored near the cannons. A stray spark from a nearby fire—which the artillerymen needed to fire their cannons—could set off an explosion.

Even worse was the fact that the gunpowder had to be mixed on the battlefield. Carrying premixed gunpowder for long distances, in a cart or on the back of a horse or a mule, jostled the compound so much that the ingredients sifted apart into separate layers of carbon, sulfur, and saltpeter. Opening many casks of separated gunpowder and remixing them would have taken too long for an army's guns to be of any use. To guarantee a steady supply of gunpowder, every army that used gunpowder had one or more men (usually civilians or mercenary soldiers) who maintained the supply of ingredients and who mixed gunpowder when it was needed. During long battles, the cannons could easily go through tons of gunpowder. And no solider relished the idea of standing next to a man whose work—conducted in the heat of battle—could literally blow up in his face.

Gunpowder's Limits

Gunpowder had proved to be a useful tool for warfare. Yet gunpowder was not powerful enough to perform much useful civilian work. Mining and construction were the only industries that had any use for explosives. And even in these fields, the tasks that gunpowder could perform were limited. The same

Early cannons used the force of a gunpowder explosion to hurl a heavy iron ball from a long tube. Although a blow from the cannonball could be deadly, the cannon's massive weight made it difficult to move.

Revolutionary War soldiers transport a cart full of gunpowder to the battlefield. Gunpowder could only be carted for short distances, as the jostling that occurred during longer trips often caused the premixed ingredients to sift apart, rendering the gunpowder useless.

amount of powder that could throw hundred-pound cannonballs through the air was virtually powerless in breaking apart rock for tunnels or in tumbling hillsides for roads. Miners and engineers had to pack hundreds of pounds of gunpowder into a single shot to have any effect. But they did not like to do this. Gunpowder was a tricky material, and there was always a danger that the miner who set off a charge would be killed in a premature blast.

Nevertheless, many miners found that it was much easier, quicker, and cheaper to replace human labor with gunpowder for some tasks. In large amounts, gunpowder could break apart large quantities of rock almost instantaneously. Until gunpowder became a tool for miners as well as soldiers, breaking rock was a hard, time-consuming process, as described by historian Donald Barr Chidsey in *Goodbye to Gunpowder:*

> A fire was built alongside a boulder in which it was hoped that ore would be found, and this was kept burning for hours on end, sometimes even for days, until the rock was heated through and through. Then cold water was dashed over the rock, and if as a result it split open, it was worked upon with hammers and picks; if not, another fire was started.

The first recorded case of blasting for minerals with black powder took place in Hungary in 1627. A German engineer named Kaspar Weindel used black powder in one of the royal mines. The explosion Weindel created broke up as much ore-bearing rock as a team of men could have produced in a month of digging. This experiment proved that blasting was more efficient than breaking rocks by hand, and helped influence miners in other nations to begin using black powder themselves. Even though gunpowder still was a fairly expensive commodity, it paid for itself in the speed with which it could bring coal, limestone, and ore-bearing rock out of the ground.

The use of gunpowder for blasting rock eventually passed beyond mining and into other areas of civil engineering, such as road and tunnel construction. In 1679, fifty-two years after its first use in Hungarian mines, workers used gunpowder to blast a tunnel for a canal in southwestern France. This was the first of many projects in which black powder was used to create waterways and roads. By 1700, gunpowder was as widely accepted as a tool of excavation as were picks and shovels. In pre-Revolutionary Connecticut, for example, gunpowder was used to exploit a large deposit of copper near the town of Simsbury. Later, the colonial government ordered workers to blast out a small room in the mine to house prisoners, creating Connecticut's Newgate Prison.

Before gunpowder became available, breaking rock was a painstaking process. Miners worked long and hard splitting rock with hammers and picks.

Gunpowder greatly speeded up the process of tunnel construction. Here, workers transport cartloads of blasted rock out of the portal of a railroad tunnel.

Gunpowder Fuses

In the century after the American Revolution, engineers used massive amounts of gunpowder to construct more than sixty canals (such as the Erie Canal in New York State) and railroads throughout what is now the eastern United States. By 1860, U.S. gunpowder manufacturers were making more than twenty-five million pounds of gunpowder a year. Yet, even with its demonstrated advantages, gunpowder still sparked a love-hate relationship with those who used it. The effort of setting up a single successful shot was a matter of close observation, detailed measurements, and luck. For one thing, until the 1830s most gunpowder blasts were set off using a straw, a hollow reed, or a goose quill filled with gunpowder and stuck in a larger mass of the explosive. The speed at which these primitive fuses burned depended mostly on how firmly the powder was packed in its little tube—a firm, tight pack meant a longer burn time, and thus more time to get away before the explosion. Because no one could ever be sure exactly how tightly the powder was packed, the men who lit the fuses did so with the knowledge that they might not be able to get away before their time ran out.

Even when the first string-type fuses came out, they were barely more reliable than the goose-quill method. Little more than cotton threads smeared with a paste of gunpowder, these first string fuses could not be depended on to burn at any particular rate. The speed at which the fuse's flame traveled depended on the amount of powder that stuck to the threads after the paste dried.

The first truly predictable fuse did not come along until 1831. A British leather merchant and inventor named William Bickford became interested in the problems created by unreliable fuses and decided to solve them. His device, the world's first "safety fuse," was simply a line of gunpowder, tightly wrapped in jute yarn and a few layers of other fabrics. When ignited, the fuse burned at a fixed rate, roughly a yard every two minutes, depending on how tightly the gunpowder in the fuse was packed. Charges of gunpowder now could be counted on to explode within a definite period after the fuse was ignited. Timing explosions still was tricky work—the fuses did not *always* burn *exactly* at the advertised rate—but now blasters had a little more control over events.

Before the advent of the safety fuse, gunpowder users could not predict how much time they had to get away before the explosion. Sudden explosions often resulted in disaster.

Workers in the Hoosac tunnel seek refuge from flying rocks during an explosion. Even with gunpowder, tunnel workers had only progressed a mile and a half after ten years of construction.

Even so, a more reliable fuse was not enough to make gunpowder a satisfying civilian explosive. It still took a huge amount of black powder to break apart even a small amount of rock. Often, there was no other way to do things than to pile casks of powder next to the wall of a mine and hope the rock was weak enough to break apart. Carefully planned blasts, where the force of the explosion was guided in a particular direction, were all but impossible. Blasting out mines and tunnels was very slow work as well. For example, construction of the nearly five-mile-long Hoosac railroad tunnel in Massachusetts was started in the mid-1800s using steam drills and black powder. Ten years later, workers had managed to penetrate a little more than a mile and a half into the side of the hill.

Just as the Western world's armies had needed a more powerful alternative to Greek fire, by the mid-1800s the world's industries found themselves in need of a more powerful alternative to gunpowder. Fortunately for them, chemists working in Europe had by this time taken an intense interest in creating new explosive compounds. And an industrialist whose family made its money making and selling explosives was about to send the world on a great technological leap forward.

■■■■■■■■■■■ CHAPTER 2

The Birth of High Explosives

Until the 1800s, the few scientists who tried to create improved explosives came up against the same obstacle. No matter what combination of chemicals they tried, they could not create an explosion more powerful than that of gunpowder. Then, in 1805, an English chemist named John Dalton developed a theory that pointed toward a new way of cooking up explosives.

English chemist John Dalton's theory about how elements combine revolutionized chemistry and paved the way for the creation of more powerful explosives.

Dalton suggested that all matter was made up of extremely tiny particles, which he called atoms. Atoms, Dalton said, were so small that people could not see them even with the strongest microscopes available. Furthermore, he said, there were different types of atoms. These different types, called elements, bonded together in groups called molecules, which in turn formed all solids, liquids, and gases in the world.

This new idea of how elements combined revolutionized chemistry. Until Dalton proposed his theory, most scientists believed that most chemicals were simply mixtures of different elements, with no physical connection between the elements. The idea that elements existed as single particles that bonded to each other overturned that long-held theory. It also allowed scientists to explore new ways to combine different elements. For most of the nineteenth century, scientists made a number of dramatic discoveries that involved previously unknown chemical combinations. Some of the most dramatic of these discoveries involved new types of explosives.

A New Look at Gunpowder

Dalton's theory helped scientists understand why gunpowder had such a weak effect on rock and how they could create more powerful explosives. The key

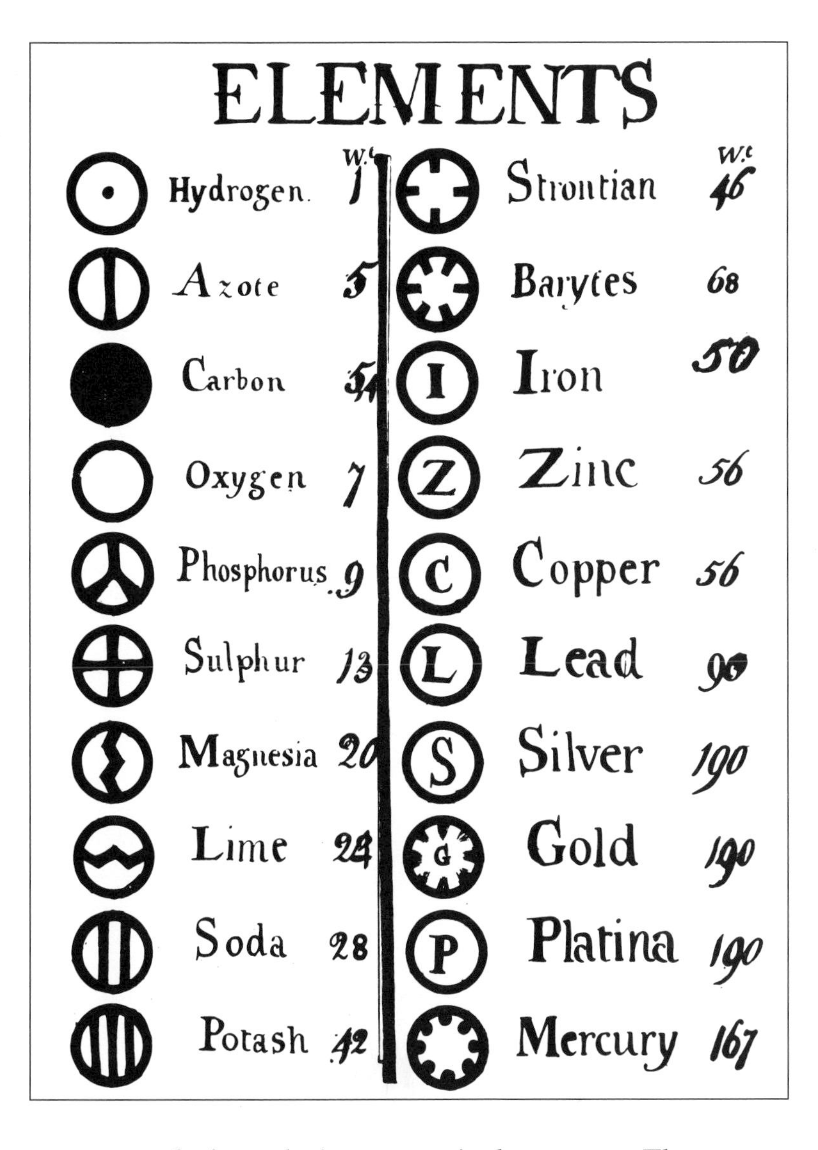

Dalton's 1803 table of elements lists the symbols and atomic weights of twenty elements. Dalton's suggestion that these elements exist as single particles that bond together contradicted the long-held theory about how elements combine.

to gunpowder's explosive nature is the saltpeter, a combination of potassium, nitrogen, and oxygen that is also known as potassium nitrate. The bonds connecting the nitrogen to the other elements are fairly weak. When heated by a flame or fuse, these bonds break apart quickly, releasing oxygen molecules that help burn the sulpher and charcoal, and keep the reaction going.

As battlefield experts had known for centuries, gunpowder was merely a physical mixture of its various ingredients. There was no molecular bond between the saltpeter, sulfur, and charcoal. Therefore, the only real interactions between the three components were a transfer of heat from the sulfur to the saltpeter and a release of heat by the oxygen. This was a fairly inefficient use of energy, because much of the heat created was used up by the sulfur and the saltpeter to keep the gunpowder burning.

A much better explosive, chemists realized, would be one in which nitrogen, oxygen, and carbon were joined together on one molecule. The right combination would create a molecule that, thanks to the nitrogen, would always be on the verge of breaking apart. The force of the ruptured chemical bonds would start the reaction going: oxygen and carbon would burn immediately, giving off the gases that did the work of the explosion. It would not be necessary to set this explosive on fire. A simple physical shock would be enough to break the bonds holding the molecule together.

The Search Begins

Scientists now had a goal to aim for: a highly powerful explosive that contained nitrogen, oxygen, and carbon on the same molecule. The first thing they had to do was to see if such a combination actually could be created in the laboratory. Most nineteenth-century research into explosives centered around compounds made with nitric acid. Nitric acid, a powerful acid that can eat through metal, contains both oxygen and nitrogen. Because of this combination, nitric acid gave scientists a starting point that already was two-thirds of the way to their goal. All they had to do now was figure out a way to add carbon to the acid.

Because carbon is one of the most common elements on earth, it was simply a matter of deciding which carbon-bearing material to use. There was just one drawback. So many things contain carbon—wood, coal, cotton, cloth, paper, chalk, sugar, leather, and much more—that it took years to mix them all with nitric acid. Most of these experiments added up to nothing more than unsightly, acidic messes on a laboratory table. However, two attempts yielded results that showed immediate promise.

Guncotton and "Blasting Oil"

The first of these successful mixtures was created by Christian Schönbein, a German chemist, in 1846. Schönbein thought that mixing nitric acid with thick cotton cloth would create a powerful yet portable explosive. The scientist proved himself right when, testing a particular mixture, he set off an explosion that yielded more energy than the

While Christian Schönbein's guncotton proved to be too powerful an explosive for use by the military, his invention led the way toward development of smokeless explosives.

Explosions from cannons using traditional gunpowder created clouds of thick, black smoke that obscured the battlefield. Smokeless explosives like guncotton held much promise for the world's armies.

same amount of gunpowder, but gave off little or no smoke. With this act, Schönbein proved that mixing nitric acid with another material—a process that became known as nitration—would yield an explosive. Schönbein's explosive, which he called guncotton, held much attraction for the world's armies. The traditional type of gunpowder used in field artillery created clouds of thick, black smoke, which came mostly from the unburned carbon in the mix. After a few volleys of artillery fire, the smoke obscured the field of battle, hiding troops from friend and enemy alike. Schönbein's guncotton turned out to be too powerful to be used as a military propellant, but it pointed the way toward the invention both of a practical form of guncotton—cordite—and to a range of smokeless powders.

The civilian world finally got its miracle explosive in 1846, a few months after the invention of guncotton. The explosive, called nitroglycerin, was developed by Ascanio Sobrero, an Italian professor of chemistry. Sobrero had noticed that many of the failures in creating nitrate-carbon explosives came from using solid compounds—wood, coal, and so forth. But, Sobrero reasoned, there are many liquid or semiliquid compounds that also are rich in carbon. Might not one of these work better? Sobrero decided to see what would happen if he added nitrate to glycerol, an oily by-product of the soap-making industry that was sold as a skin lotion. Glycerol was cheap and, because of a nearby soap factory, easy for Sobrero to obtain. The result of the experiment was an opaque yellow liquid that, at

Despite the inventor's warnings about the dangers of nitroglycerin, scientists and engineers eagerly began their own experiments with the liquid explosive. Even the U.S. government conducted tests comparing gunpowder and nitroglycerin, as depicted by this illustration.

first, seemed doubtful as an explosive. When Sobrero heated a tiny amount of the mixture to see how it would react, it merely gave off a thin, sickly sweet vapor that gave him a splitting headache.

Then the chemist decided to take one extra step. He soaked a piece of paper with just a few drops of the nitrate-glycerol mixture—called nitroglycerin—and hit it with a hammer to judge how sensitive it was to a simple shock. The resulting explosion may not have flung the hammer from Sobrero's grasp, but it certainly, according to witnesses, rattled the windows of his laboratory.

In the interest of science, Sobrero felt compelled to publish the result of his experiment. However, he was so alarmed at the destructive potential of nitroglycerin that he warned strongly against its use, and abandoned his research. But other scientists, as well as enterprising miners and engineers, realized that Sobrero had just given the world a new tool for large-scale excavations. They were eager to run their own tests on the Italian chemist's invention.

Unfortunately, for the next fifteen years nobody could figure out how to control the oily explosive. Regular gunpowder-based fuses were no good. As Sobrero had shown, the only result of heating nitroglycerin was a bad headache, and all a conventional gunpowder fuse did was carry a flame to the explosive. Some scientists and inventors tried mixing nitroglycerin with gunpowder, hoping to make the powder stronger. This attempt was only moderately successful, as it gave only a slight boost to the power of plain gunpowder. And no one was about to try using Sobrero's method of whacking nitroglycerin with a hammer. Clearly, the practical use of this powerful explosive was beyond the known methods of setting off explosives.

Nobel's Igniter

Then, in 1861, a Swedish chemist and industrialist named Alfred Nobel came up with an idea that would usher in a new era of explosives engineering. Nobel had been part of the explosives industry for nearly his entire life. Soon after his birth in 1833, his father Immanuel moved the Nobel family to the Russian city of St. Petersburg and went into business making gunpowder-based land and sea mines for the imperial Russian military. When Russia canceled most of its contracts in 1856, the Nobels moved back to Sweden, where Immanuel continued his work.

Immanuel Nobel was one of those who tried combining gunpowder with nitroglycerin, but with little success. Although the gunpowder-nitroglycerin mixture was stable, it was only slightly more powerful than plain gunpowder.

Alfred Nobel's igniter enabled scientists and commercial blasters to control the power of nitroglycerin, ushering in a new era of explosives engineering.

His son Alfred, the youngest of his five children, had been working in the family business for roughly ten years by the time the family returned to Sweden. Alfred saw possibilities in nitroglycerin despite its poor performance as a booster for gunpowder. Alfred thought that the key to controlling nitroglycerin was precisely in what made it so dangerous—its sensitivity to all but the mildest shocks.

One day in 1863, Alfred realized that all the work that he and other scientists had done was focused in exactly the wrong area. Instead of getting nitroglycerin to explode by itself, he reasoned, there should be a way to use some other force to shatter the bonds holding its molecules together. As soon as this thought hit him, the answer to the problem of controlling nitroglycerin fell into place. From his knowledge of how atoms form molecules, the theory first proposed by John Dalton, Nobel knew that the nitrogen-oxygen-carbon compound was already inclined to break apart. Gunpowder fuses, which only carried heat to an explosive charge, did not have the power to break apart the nitroglycerin molecules. But a small, sealed charge of gunpowder, set off inside or next to a larger amount of nitroglycerin, would give the liquid explosive the shock it needed to explode, Nobel thought.

With this step of reasoning, Nobel created a principle that has been used in every type of modern explosive device—the primary detonator, or primer. This idea also put the Nobel family at the forefront of the world's explosives industry. Soon after perfecting his gunpowder primer charge, Nobel began selling it under the name "Nobel's Igniter." He combined the igniter with containers of nitroglycerin as an easily

NOBEL'S IGNITER

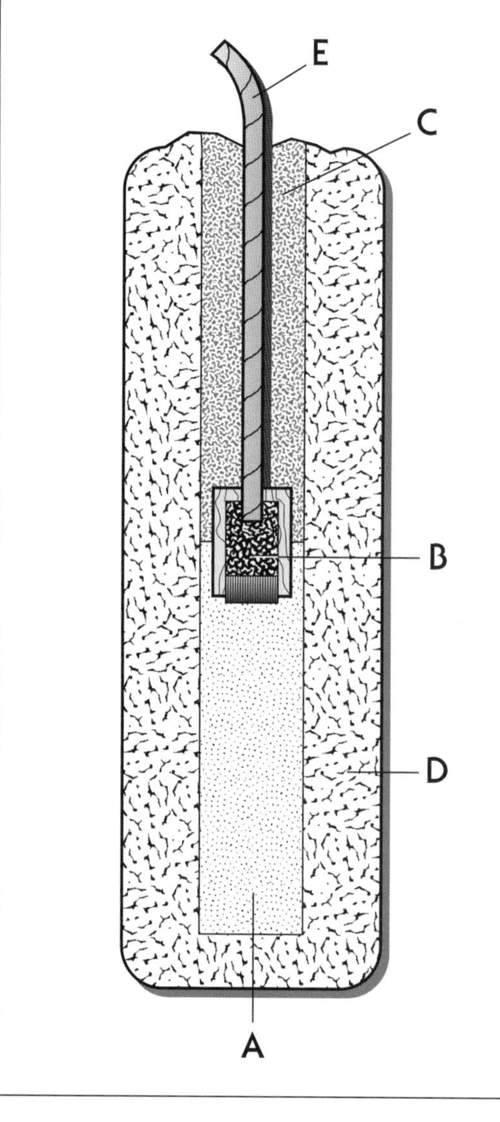

Alfred Nobel's igniter was a milestone in the history of high explosives. It was the first practical device for controlling the power of nitroglycerin. It also introduced the technique of primary detonation to the field of high explosives. With primary detonation, a small, easily controlled explosive charge detonates a larger charge. This technique improved safety and served as a model for later explosives.

Nobel's device contained a measured amount of nitroglycerin (A), a small metal cartridge filled with gunpowder (B), and a cork or plug of an inert material (C). The gunpowder cartridge was the primer, the device that safely detonated the main charge—in this case, the nitroglycerin. All these materials were held inside a small, well-insulated container (D). The primer was ignited by lighting a safety fuse (E). The length of this fuse determined how much time passed before the nitroglycerin exploded.

Nobel's igniter did not entirely do away with the risks of using nitroglycerin. But because the igniter offered greater safety and control than had previously existed, Nobel and others could move ahead in experiments that would lead to dynamite and other explosives.

used commercial explosive. Nobel eventually replaced the gunpowder primer with a small metal cylinder filled with mercury fulminate, an explosive mixture of mercury, carbon, nitrogen, and oxygen that had been discovered in the 1820s. In large quantities, mercury fulminate was too powerful for commercial or military use. But in the small capsule Nobel devised—the world's first blasting cap—the explosive was just powerful enough to set off the nitroglycerin.

Nobel opened igniter factories in Sweden and Germany to try to handle the small but growing demand for his invention, which was given the nickname "blasting oil." However, the brief popularity surrounding Nobel's blasting oil soon vanished in the wake of numerous accidental explosions caused by the unstable liquid. Sobrero had learned firsthand about nitroglycerin's sensitivity. But until Nobel began mass-producing the explosive, nobody knew just how sensitive it really was. The bonds tying the atoms of nitrogen, oxygen, and carbon to each other were the chemical equivalent of a hair trigger. Any shock rougher than a slight jostling shattered these bonds, releasing the energy that held the nitroglycerin molecules together.

The risks of making nitroglycerin were far greater than the risks of making gunpowder. With gunpowder, there was always a risk that a spark or a stray flame might set off an explosion. With nitroglycerin, even a sharp tap of a stirring rod against a kettle could lead to disaster. Ironically, the Nobel family became one of the first victims of the chemical's volatile nature. In September 1864, the Nobel nitroglycerin factory in Stockholm, Sweden, suddenly exploded, killing three workers, a visitor to the factory, and Alfred's brother Emil, who had been supervising that day's shift. Over the next three or four years, hundreds of people were killed in other accidental explosions around the world: on a ship carrying a load of nitroglycerin in rough seas off the Panama coast; in a San Francisco warehouse where crates of the explosive were being stored; in New York City; in Sydney, Australia; even in another Nobel factory in Sweden. It seemed that no one was safe when nitroglycerin was around. People began to think that even a sneeze was enough to set it off. To ward off any future disasters, many nations either banned the blasting oil altogether or placed heavy restrictions on how and where it could be made, shipped, and used. Nobel's gift to the world had proved itself to be a curse in disguise, and the world wanted nothing to do with it.

The aftermath of an 1889 gunpowder explosion at a Du Pont factory. Accidental explosions at explosives-manufacturing plants claimed many lives.

The volatility of nitroglycerin made it extremely dangerous to manufacture. The frequency of accidental explosions prompted Nobel to publish a method for its safe handling and transportation.

Nobel's Powerful Solution

Alfred Nobel remained convinced that, despite its dangers, the power of nitroglycerin was a great boon to humanity. He also honestly believed that the many accidental explosions resulted from carelessness, rather than from any particular characteristics of the explosive. Nobel had even published a method for the safe handling and transportation of nitroglycerin, and he believed the substance could not be blamed if people chose to ignore his advice. Testifying before a Swedish court in defense of nitroglycerin, Nobel said:

> It cannot be expected that an explosive substance should come into general use without waste of life. A simple reference to statistics will show that the use of firearms for play . . . [creates] incomparably more accidents than this substance, which is a great and valuable agent for the development of our mineral wealth.

Still, Nobel realized that if nitroglycerin was to be given a chance to prove its worth, he would have to find a way to counteract its instability. He also knew that if he did not rid nitroglycerin of its dangerous image, his family would be put out of business. But how could he stabilize the oil while keeping its destructive power intact?

At first, Nobel tried adding thickeners that would change nitroglycerin to a solid, stable form. He pinned most of his hopes on wood alcohol, thinking that the alcohol would make the solution congeal. But that method did not work, nor did other thickeners he tried. Eventually, Nobel decided he would have to find some solid material that could act as a buffer by absorbing nitroglycerin. He tried a host of different solids, including coal, chalk, paper, and even brick dust. But none of these materials worked.

Finally, in 1867 he found something that would both absorb nitroglycerin and keep it stable until it was made to explode. The material he used was kieselguhr, a silicon-rich, sandy soil that soaked up the blasting oil much like a sponge soaks up water. Nobel found that he could hit bricks made of the oil-soaked kieselguhr, and even melt pieces of it in a flame, with no risk of an explosion. Yet, when he paired a small piece of the blasting oil-soaked kieselguhr with a blasting cap, he could release nearly all of the energy of the trapped nitroglycerin. And most importantly, the energy contained in one pound of the new material was equal to that of more than twenty pounds of black powder.

Such a powerful explosive, Nobel felt, deserved an equally powerful name. Searching through his knowledge of languages, Nobel remembered that the Greek word for power is *dynamis*. Pairing this word with an ending that means both "product of" and "rock," Nobel gave his new solid explosive a name that, for the next century, came to symbolize all explosives: dynamite.

The mixing house at Nobel's dynamite factory. By mixing nitroglycerin with kieselguhr, Nobel invented dynamite, a solid explosive whose invention heralded the era of high explosives.

An Explosive New Era

The invention of dynamite in 1867, even more than Nobel's development of the blasting cap, marked the beginning of the era of high explosives. For the first time, the public had access to nitroglycerin in a form that was easier to use and safer to handle than gunpowder. The use of gunpowder declined slowly but steadily as more people began using the solid mix of kieselguhr and nitroglycerin. Soon, the only use for gunpowder was as a propellant for firearms and fireworks. And even then, it was gradually replaced by more powerful, smokeless powders.

With the creation of nitroglycerin and dynamite, Sobrero and Nobel also had created an entirely new class of explosives. Gunpowder mixed chemicals that merely burned fast, a process called deflagration. A trail of gunpowder was consumed at a rate of, at most, a few feet per second, and created a gas pressure of fifty thousand pounds per square inch (psi). Nitroglycerin and dynamite were consumed at a rate of up to five miles per second, and generated more than one million psi. And instead of burning, they literally flashed from a liquid or a solid to a hot, high-pressure gas, a process that became known as detonation. Deflagrating explosives, because they merely burned and put out such comparatively low pressure, became known as low explosives. Detonating explosives thus became known as high explosives.

In general, explosives that burned slower than five thousand feet per second and generated pressures of less than fifty thousand psi were defined as deflagrants. Explosives that were consumed faster and generated higher pressures were said to have made a deflagration-to-detonation transition, or D.D.T. This definition was not an absolute measurement of the explosive's power, however. Gunpowder could be made to perform almost like a high explosive, as miners had done during the two hundred years before dynamite's invention. By confining the blast in a tight space, miners were able to artificially raise the pressure generated by the deflagrating gunpowder. This confinement made the gunpowder's gases work harder, though without anything approaching the force of a high-explosive blast.

Despite his success with dynamite, however, Nobel was not satisfied with simply producing it as it was. He conducted numerous tests on the combination of sandy soil and blasting oil to see if it was giving off the largest possible explosion. He set off charges of liquid nitroglycerin and charges of dynamite that contained the same amount of the blasting oil, then compared the force of the two explosions. He found that the dynamite explosion actually was less powerful than that of the nitroglycerin. Nobel realized that rather than simply evaporating when a blast was set off, the kieselguhr actually absorbed some of the heat of the explosion. This effect reduced the speed at which the nitroglycerin molecules broke apart, and effectively muffled a part of the explosion.

The problem was relatively simple to solve, however. All Nobel had to do

Coal miners wait for the gunpowder blast. Classified as a low explosive, gunpowder burned slower and generated less pressure than dynamite, a high explosive.

DETONATION WAVE

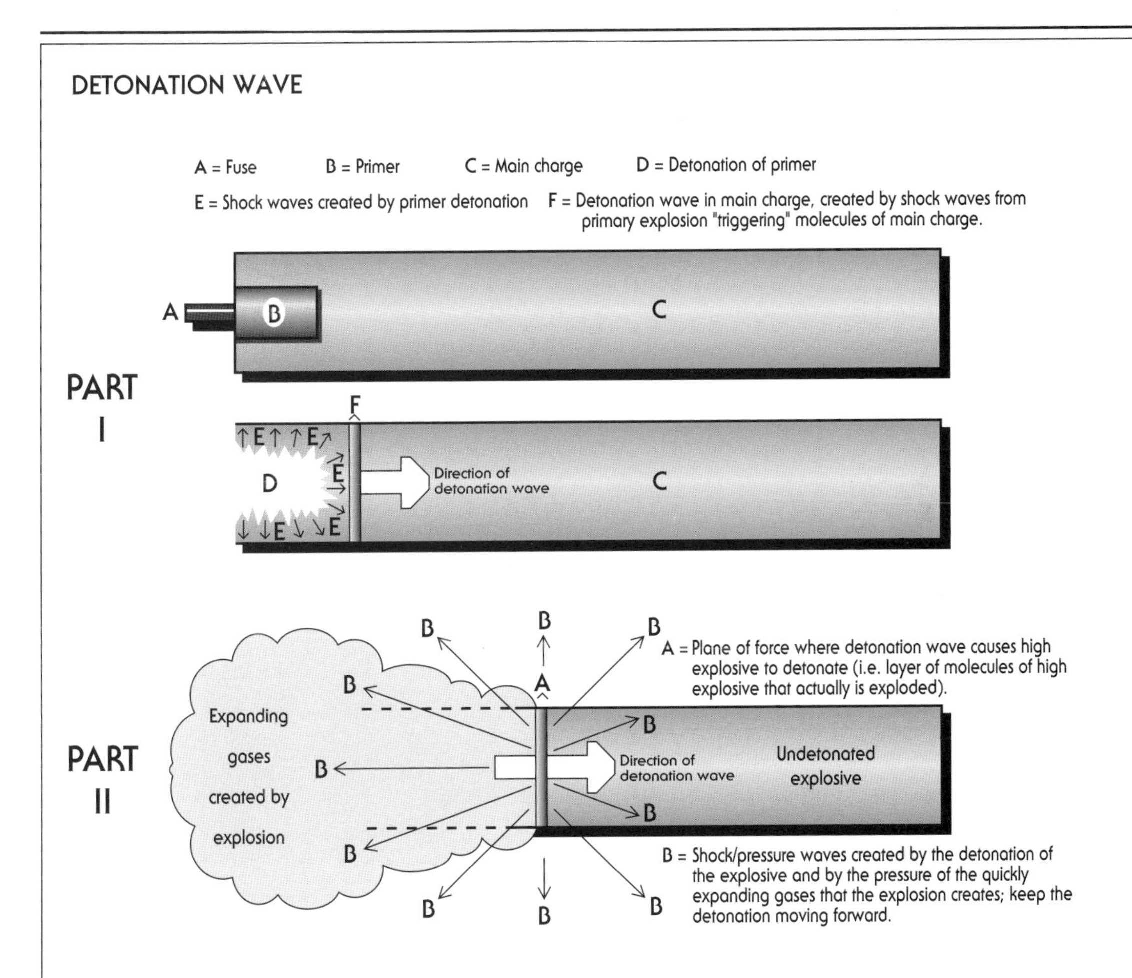

PART I:

The detonation of the primer only affects one part of the main charge—the part near the primer. In order for the entire main charge to explode, some force has to travel through the charge, forcing all its molecules to detonate. That force is the detonation wave.

A typical explosive charge has three parts: the fuse (A), the primer (B), and the main charge (C). When the primer, ignited by the fuse, detonates (D), it sends out shock waves in all directions (E). Some of these shock waves travel towards the center of the main charge. The pressure and the heat they create violently rupture the molecules of the main charge, starting a chain reaction that passes through the length of the charge.

PART II:

In order to more easily understand this reaction, scientists think of the chain reaction as a vertical plane of force (A)—a wave of detonation—that rams its way through the explosive. The actual rupturing of molecules starts inside this plane, and the generation of gas takes place just behind the plane. The explosion's shock waves and the pressure of the expanding gases (B) help drive the detonation wave through the entire mass of explosive.

The packing house at Nobel's dynamite factory. Nobel began packing dynamite in thick cardboard tubes and wrapping it in waxed paper to keep it from absorbing water.

was find a suitable replacement for the kieselguhr. After testing a number of materials, he decided to use a combination of wood pulp and sodium nitrate, which itself burned and added heat to the explosion. Nobel named this mixture "dope," and began grading dynamite according to the percentage of nitroglycerin that was in the mix with the dope. To protect the explosive from water, which tended to leach out the nitroglycerin, Nobel began packing dynamite in thick cardboard tubes and wrapping them in a couple of layers of waxed paper.

Nobel further improved the power and lowered the cost of dynamite by substituting ammonium nitrate for some of the nitroglycerin. In 1867, the same year Nobel invented his first dynamite, two other Swedish chemists patented a method that used ammonium nitrate—a mixture of nitrogen, oxygen, and the extremely flammable element hydrogen—to make a number of powerful explosives. Nobel bought the patent from his countrymen—known only as Ohlsson and Norrbein—and applied its techniques to his own explosives.

Despite being slightly less powerful than nitroglycerin, ammonium nitrate had the benefit of being more difficult to detonate. This added resistance, or insensitivity, made the new form of dynamite safer to handle. With these two forms of improved dynamite, Nobel and the world had the full power of nitroglycerin under their control.

CHAPTER 3

An Explosion of Explosives

In all its forms, dynamite was the first truly practical high explosive to be developed in the nineteenth century. But it was not the only one. While Alfred Nobel was creating and improving his invention, many other scientists were trying to repeat his success with their own nitrogen-carbon compounds. And still others were searching for explosives that did not require the dangerous methods of mixing acids and solids needed to make nitroglycerin and guncotton. Between 1870 and 1910, more than twenty different types of explosives were developed, and many more ways were found to combine them for greater effect. The work of these scientists was so detailed and used such a wide range of elements that most of the explosives used today were developed by the end of the first decade of the twentieth century.

For a time, however, dynamite was really the only high explosive in existence. Dynamite was so powerful and had so many uses that there did not seem to be a need for any other. For example, dynamite dramatically increased the amount of metals, coal, and other materials produced by the world's mines. A few pounds of dynamite, with one explosion, could excavate as much rock as a team of workers, chipping away with picks and hammers, could dig through in a month. With dynamite, the only physical labor needed was to place the explosive in holes drilled in the rock and to carry the shattered ore

Miners load shattered coal after a dynamite blast. Dynamite dramatically increased mining production. A single dynamite explosion could break as much rock as a team of workers with picks and hammers could dig through in a month.

to the surface. Mining costs dropped to the point that miners were able to exploit smaller deposits, ones that had not been worth enough to pursue with older methods. With dynamite, virtually any deposit, no matter how small, could be made to bring in a profit.

Dynamite Economics

Increasing the outputs of the world's mines had a cascading effect on other industries. Until the late 1800s, the most advanced building materials available were bricks and mortar. Although these materials could be made into sturdy buildings, brick buildings could not be built very high. Ten stories were as high as any building got, and then only with careful engineering. Any higher and the bricks might crumble under their own weight. With the appearance of dynamite, however, came increased output from quarries of limestone—a key ingredient of cement—and of gravel, which is mixed with cement to form concrete. Concrete, when mixed correctly, is able to support far heavier loads than bricks and mortar can. Better still, it can be poured into any shape and allowed to harden in place. This way, it can be used to make single-piece walls and floors that are stronger than they would be if they were made out of thousands of small bricks. The concrete revolution, spurred by dynamite, in turn spurred the architectural revolutions that gave the United States its dams and its highway system, and that gave the world skyscrapers.

Dynamite-enhanced mining and construction also gave the world a new metals industry. Until the turn of the century, aluminum was a more valuable metal than either gold or platinum. The basic aluminum-bearing ore, bauxite—a combination of aluminum, oxygen, and hydrogen—is one of the most common elements on earth. Yet separating aluminum from bauxite demands huge amounts of heat, amounts that simply could not be generated for long by furnaces built before the mid-1800s. It took a fortune in coal to melt out a usable amount of aluminum from bauxite. But with the new concrete industry came the capacity to build huge dams that doubled as hydroelectric power plants, housing gigantic turbine generators inside sturdy concrete walls. These power plants gave out more than enough cheap electricity to run electric ovens that could smelt large quantities of aluminum, making the formerly priceless metal as common and as inexpensive as paper.

A Pennsylvania limestone quarry. Dynamite greatly increased output from quarries of limestone and gravel, key ingredients in cement and concrete.

The importance of dynamite was felt almost immediately. The concrete revolution, spurred by dynamite, in turn gave rise to the architectural revolutions that gave the United States its dams (below) and highway system (right).

Railroads, ocean shipping lines, and even river traders benefited from dynamite as well. One of the first large-scale construction projects to use dynamite in the United States was the mile-long Musconetcong railroad tunnel near Philipsburg, New Jersey. Starting in 1872, engineers used dynamite to blast out more than 135 feet of granite in a single month. The Musconetcong tunnel was completed in less than four years, a phenomenal feat compared with the construction of the Hoosac tunnel in Massachusetts, where crews using steam drills and gunpowder progressed only a mile and a half in ten years. Soon, work was starting on what would become hundreds of dynamite-cleared tunnels around the world. These tunnels allowed trains to travel on a straighter line between cities, keeping them from having to snake around mountains and hills. They also traveled on flatter, safer paths, built on land flattened using dynamite and set on beds of gravel excavated by dynamite. Dynamite wrapped in waterproof paper or fabric was used to blow up underwater

BOREHOLES

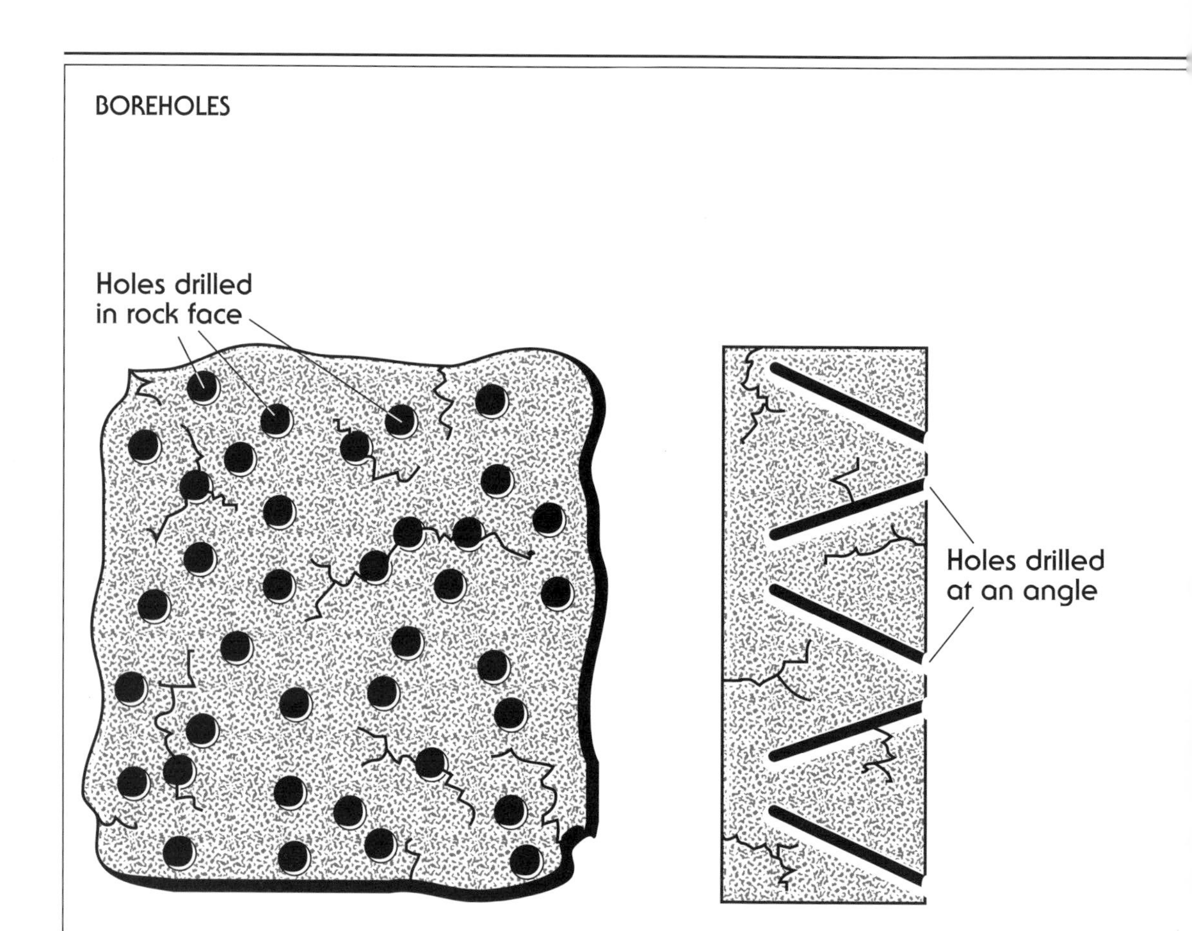

Blasting rock in mines and tunnels is a very detailed engineering feat. To blast tunnels and mines through rock, miners and excavators have to drill, or bore, holes in the rock. These holes, called boreholes, hold the explosives that break up the rock. The boreholes are made with a high-powered drill mounted on a framework called a jumbo. The jumbo guides the workers and supports the surrounding mine or tunnel walls. The blasting area, or rock face, of the mine or tunnel may contain as many as two hundred holes, each about six inches wide and up to twenty feet deep.

The placement and the angle of the boreholes is important. The pattern of the holes determines where the rock breaks. Usually they are drilled to fracture the rock at its weakest points. Angled holes are preferred over holes drilled straight into the rock because angled holes break up more rock.

An explosive charge is placed in each borehole. Each charge is then wired to a detonator. The detonator explodes the charges in a carefully timed series of blasts. The first blasts blow a hole in the middle of the rock face. Broken rock from the next explosions falls into the hole and then onto the floor of the tunnel, making a new rock face for the next series of blasts.

obstructions in the mouths of rivers and harbors. Dynamited channels in the middle of harbors also allowed bigger ships, which sat lower in the water, to make use of many ports without fear of scraping their keels or grounding themselves. And dynamited rocks, hauled from mines and construction sites, were used to make sheltering breakwaters that protected these improved harbors.

Tunneling Through the Alps

Dynamite's ability to speed up work on major construction projects was demonstrated shortly after its invention in 1867, during the construction of the first tunnel through the Alps. In the 1850s, work began on a railroad tunnel beneath Mont-Cenis, an Alpine mountain that straddled the border between northern Italy and southern France. The tunnel was designed to increase commercial trade between Italy and France, a trade that had for centuries been restricted to small teams of wagons that crossed treacherous passes over the Alps. By using air-powered drills and long cases of gunpowder, the tunnel's builders said, they could have the tunnel ready for use in forty years.

The Mont-Cenis tunnel project was supposed to revolutionize the way tunnels were constructed. The high-pressure air lines that powered the drills in the mines also were supposed to clear the air of the thick smoke generated by the gunpowder charges. By clearing out the working area, or heading, the attack crew—the workers who set the charges and who cleared rock with horse-drawn wagons—would be able to work quickly and easily.

As often happens, however, things did not go according to their designers'

Dynamite became a useful tool for moving underwater obstructions in the mouths of rivers and harbors. (Left) The blast at Ripple Rock, a massive underwater mountain and one of the world's worst navigation hazards. (Right) The tunnel under Ripple Rock, used by workers to set up the explosions and cart out the broken rock.

plans. In *Adventure Underground: The Story of the World's Great Tunnels,* author Joseph Gies described how the Mont-Cenis project actually looked:

> The scene inside the heading was murkily fantastic. The walls dripped continuously. The oil lamps hardly penetrated the black smoke which virtually filled the entire heading, from face to portal. The jets from . . . compressed-air hoses cleared only the area directly in front of the [blasting] face, to permit the attack crew to operate. From there on back the smoke simply hung in the tunnel. The laborers breathed it, half-choked from it, often were overcome by it. It was impossible to see more than a few feet. One English visitor [in 1866] reported, "I could not see a lamp on the opposite side. The horses and wagons passed, but I could not see them. This continued up to within one hundred yards of the end, when a light could be seen for twenty yards. . . . I was in about an hour, and when I came out I spit as black as though I had dined on lampblack."

Then, in 1867, Nobel began selling dynamite. As soon as the news of the new explosive reached the engineer in charge of the tunnel, Germain Sommeiller, he ordered that it be used instead of the inefficient, smoky black powder. Dynamite more than doubled the speed of the tunnel's construction, because it broke apart more rock more quickly than did the black powder. The tunnel, which everyone had thought would not be in operation until the turn of the century, was in operation by 1871. But dynamite offered benefits other than a faster rate of blasting. Gies reports:

> It also made the atmosphere inside the tunnel considerably less eerie and poisonous. . . . The scene inside the heading was no longer the choking gloom of 1866; though hardly as bright as daylight, one could see, at least, by the smoking kerosine lamps. Of course the walls still dripped [with water from nearby underground streams], and occasionally even spurted, and the air remained inadequate. But the tunnel was livable.

The pace of construction on the Mont-Cenis railroad tunnel project more than doubled when workers began using dynamite instead of black powder.

The benefits of using dynamite increased the amount of work that construction crews could do. The end result of this burst of activity was a more efficient and more economical system of sea and land transportation. Cargo and passenger ships could be assured of safe harbors that offered virtually no natural hazards to navigation. Railroads could cut the time it took to travel from one city to another, and at the same time cut both their operating costs and their shipping and ticket prices. The dynamited tunnels and flattened rail beds also helped protect the trains from some of the hazards of nature, such as avalanches and sinkholes. And with transportation, especially overland transportation, suddenly so practical, more people were able to travel to and settle in once distant areas.

The size of the explosions used in this work could be quite spectacular. In 1885, in one of the world's first underwater demolition jobs, more than 288,000 pounds of dynamite was used to demolish a line of shoals, or sandbars, in New York City's East River. The shoals blocked traffic between Long Island Sound and Upper New York Bay. The explosion sent a column of water more than three hundred feet into the air, in addition to demolishing the shoals.

Private citizens also found that dynamite could make some of their tasks easier. Farmers, for instance, found dynamite useful for tasks that otherwise would have taken hours of backbreaking labor. They used dynamite to blast out deeply rooted tree stumps from their fields. They used dynamite to blast out irrigation ditches and canals for their fields. And in California, while the Los Angeles Department of Water and Power was blasting out the Owen's Valley waterway in the early 1930s, farmers and other residents of the valley who objected to the big city's intrusion used dynamite to sabotage the work.

There were just as many strange or foolish uses for dynamite as there were revolutionary ones. As early as the 1880s, dynamite fishing became a popular, if unsporting and illegal, way to bring in a big haul of fish. A stick of dynamite set up a huge wave of pressure underwater that killed fish for a few feet around the blast. All that the so-called fisherman had to do was scoop up the dead fish as they floated to the surface. As early as the 1880s, people jokingly referred to dynamite as "Atlas spinners" or "Hercules night crawlers," Atlas and Hercules being the names of two big U.S. explosives companies.

Explosive Research

As beneficial as dynamite was, it had a number of fairly severe drawbacks. One of these was the headaches caused by nitroglycerin. Absorbed either through the skin or by inhaling its fumes, nitroglycerin also caused reactions like those caused by hay fever: swollen sinuses; bloodshot, watery eyes; and, on occasion, clogged lungs. Another even more serious problem was that nitroglycerin

Workers prepare for the huge 1885 dynamite explosion under New York City's East River, one of the world's first underwater demolition projects.

freezes at a fairly warm temperature—fifty-five degrees Fahrenheit, as opposed to thirty-two degrees for water. In a small way, this high freezing temperature was a benefit. Because frozen nitroglycerin is very stable, it did not explode when dropped. But nitroglycerin, even when mixed in dynamite, became very unstable while it was thawing out. Miners or engineers who were in a hurry often tried to melt dynamite near furnaces, in ovens, or in the embers of cooking fires. The result often was an unexpected explosion that killed or maimed anyone who was standing nearby.

These characteristics did not deter people from buying dynamite. On the contrary, the worldwide demand for dynamite brought Alfred Nobel a great deal of wealth, making the Swedish chemist one of the world's richest men. But dynamite's flaws encouraged many private scientists and chemical companies alike to seek out new explosives that were even safer than the doped nitroglycerin.

Nobel himself was one of the leaders in the quest for better high explosives. Some of his innovations, like ammonium nitrate dynamite, built on his earlier work. Others resulted from luck. One day in 1875, while working in his laboratory, Nobel accidentally cut his finger and covered it with collodion, a jellylike substance that used cellulose nitrate (much like Christian Schönbein's guncotton) as an antiseptic. That night, the chemist was unable to sleep because of the annoying pain in his finger caused by the collodion. Pondering the substance that was causing his pain, Nobel wondered if combining cellulose nitrate with nitroglycerin would make an effective explosive. Even though it was 4:00 A.M., he went to his lab and began working with the two compounds.

The result of his work was blasting gelatin, a semiliquid explosive that was an early forerunner of today's plastic explosives. Blasting gelatin solved or reduced many of the problems of dynamite. It was slightly more powerful than the solid explosive, it had a lower freez-

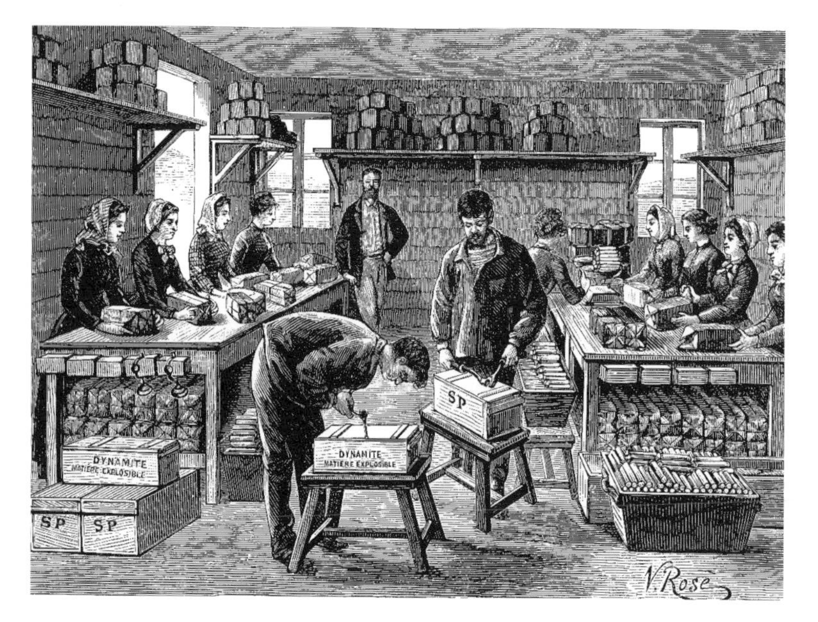

Despite the hazards of working with dynamite, demand for the new explosive quickly grew, with dynamite factories springing up all over the world.

HOW DYNAMITE IS MANUFACTURED

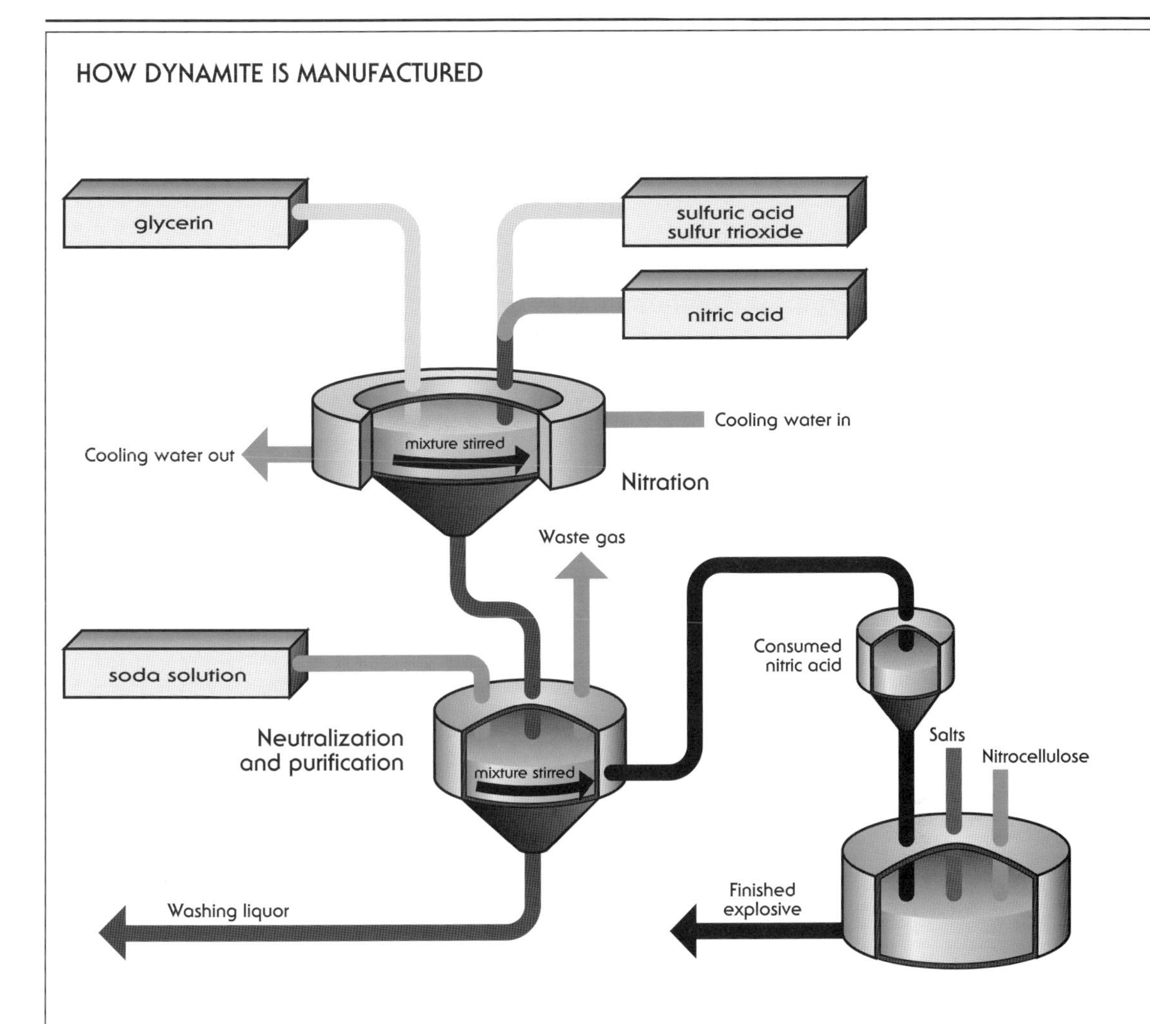

In the early days of the dynamite era, dynamite factories were primitive operations. It was not uncommon to see two workers in one building making nitroglycerin by mixing nitric acid, sulfuric acid, and glycerin in earthenware jugs using wooden ladles, while workers in another building slowly mixed the finished nitroglycerin with various other materials to form dynamite. With conditions like these, accidental explosions were inevitable, and it was truly remarkable for a dynamite factory to *not* suffer at least one explosion.

Modern dynamite factories are very different from their early counterparts. The manufacturing process is almost entirely automated, with workers controlling each step of the process from reinforced bunkers. Nitroglycerin is formed by mixing glycerin, sulfuric acid, and nitric acid in large metal vats that are surrounded by cooling water jackets. The nitroglycerin is slowly pumped to another vat, where the uncombined acids are chemically neutralized and filtered out of the mix, partly in the form of waste gas. The purified nitroglycerin then is pumped through two more vats, where it is mixed with the various fillers that turn it into dynamite.

ing temperature, and it gave off less of the headache-causing vapors. It could also be poured into blast holes, filling them more tightly and giving a more effective blast than could the rigid sticks of dynamite, which often left gaps in the blast holes.

Nitroglycerin and cellulose nitrate were not the only explosive compounds available, though. Much of the work of other scientists focused on different chemical combinations, and even different methods of making explosives. One of these scientists, an English chemist named Hermann Sprengel, thought that carrying explosives around in a ready-to-blast form was foolish. A more sensible and safer method, he thought, would be to transport the separate ingredients to the blast site and combine them just before they were needed.

Sprengel developed a method of combining solid substances that contained nitrogen, such as solid nitric acid, with burnable liquids like benzine. Because the explosive did not really become an explosive until the two chemicals were mixed, moving the chemicals about was much safer. And because no nitroglycerin was involved, there were no problems with either headaches or freezing. These mix-on-site explosives, which became known as Sprengel explosives, were not widely used in the United States, where dynamite had become the public and industry favorite. But because they were so safe to transport, Sprengel explosives became very popular in Europe.

Later, in 1895, a German chemist named Carl von Linde developed a Sprengel-type explosive that used bags of soot soaked in liquid oxygen. This combination, known as LOX, had the double advantage of being more powerful and less expensive to make in large quantities than dynamite. In the United States, LOX was most widely used in a number of open-pit coal mines, which had such a large need for explosives that it was cheaper to use LOX than dynamite. However, more powerful explosives were later developed, pushing LOX out of the blasting picture.

Nobel, the "Merchant of Death"

In their first decades of existence, dynamite and other high explosives had been used for a great deal of beneficial work. Explosives had been used to establish roadways and railways, to create new industries, and to provide the coal and other resources that powered the new burst of growth in nations around the world. Despite the good high explosives had accomplished, many people were worried about their potential for use as tools of destruction. After all, compounds powerful enough to move mountains might also be used by one nation to destroy another. The time was not far off, many people suspected, when explosives would be adapted for military use. And they were right. Before long, the world's military powers had found many uses for explosives, prompting much criticism where once there was only praise.

Much of the criticism was directed at Nobel. One day in 1888, soon after the death of his older brother Ludvig, Nobel read an obituary published in a newspaper that mistakenly identified Alfred as the one who had died. The article called him a "merchant of death" and blamed him for increasing the power of the world's armies.

Work on the last mile of the Pacific Railroad. Although the new explosives proved invaluable for railroad and tunnel construction, many people feared their use as tools of destruction.

Nobel became depressed over seeing this description of his work. He had always seen his explosives as tools of progress and civilization, not as weapons of destruction and death. Even when he was working to make more efficient types of smokeless gunpowder, which was only used by the military, he defended his work as having an ultimately peaceful purpose. In a letter to a friend who was involved in a worldwide peace movement during the 1890s, Nobel wrote:

> My factories may make an end of war sooner than your congresses . . . because the day that two armies have the capacity to annihilate each other within a few seconds, it is then likely that all civilized nations will turn their backs on warfare.

Despite his hope that such great weapons of destruction would convince the world to abandon warfare, Nobel also believed that he had to help create as strong a climate for peace as his inventions had created for war. This resolution led him to establish in his will the funds and the methods for awarding the now-famous Nobel Prizes for science, literature, and peace.

■ ■ ■ ■ ■ ■ ■ ■ ■ ■ ■ CHAPTER **4**

Explosives and the Military

The creation of high explosives was more than a boon to miners and farmers. As many people, including Alfred Nobel, feared would happen, the world's military powers saw explosives as a means to victory on the battlefield. While much nineteenth-century explosives research sought commercial explosives, many researchers also sought ways to adapt explosives to the needs of the world's armies and navies.

Not all of these explosives were used right away, however. Some were so hard to control that they were not used until a decade or more after their discovery.

A U.S. Civil War military arsenal. During the nineteenth century the world's military powers began looking for ways to harness the power of high explosives to achieve victory on the battlefield.

Among these military explosives were hexogen, or RDX, which was developed in 1899, and pentaerythritol tetranitrate, or PETN, which was invented in 1901. These explosives both were as powerful as, if not more than, straight liquid nitroglycerin. At the same time, though, they were extremely sensitive to shocks and rough handling. They were even more dangerous to transport and use than was liquid nitroglycerin. Because they were so powerful and so unstable, they were judged to be too dangerous for anyone to use. Until a way was found to stabilize them, RDX and PETN were merely interesting sidelights in the explosives field. Most of the attention at the turn of the century was on a third explosive that would become one of the world's most widely used weapons.

TNT

Of all the explosives developed around the turn of the century, perhaps the most well-known (aside from dynamite) was trinitrotoluene, or TNT. Its development as an explosive in 1904 was perhaps one of the strangest events in the history of explosives. TNT actually was discovered in 1863, about the same time as the invention of dynamite, by J. Wilbrand, a German chemist. Like most other chemists who were working with explosives, Wilbrand had been working with ways to mix nitric acid with a carbon-bearing chemical. Wilbrand had decided to use toluene, a combination of carbon and hydrogen, figuring that the highly flammable hydrogen would add extra power to the explosion.

Wilbrand chose well. TNT seemed to be a perfect explosive—it gave a powerful blast, but was relatively resistant to sudden shocks. It could only be detonated using a powerful blasting cap, such as Nobel had devised for dynamite. Even though TNT was a solid explosive, it was so stable that it could be melted and poured without risking an explosion. The temperature at which TNT melted, Wilbrand and other researchers found, was far cooler than the temperature at which its molecular bonds broke apart. This characteristic attracted a great deal of interest from munitions makers and military officers; an explosive that could be safely melted could be poured directly into artillery shells, creating powerful explosive warheads.

The only problem with TNT was that, until the turn of the century, making it in the quantities needed for civilian and military work was far too expensive. TNT required huge amounts of toluene, which was distilled from coal tar or crude oil. In the 1860s, there was no worldwide oil industry that could provide the toluene needed for large-scale weapons production. So for a time, TNT was set aside as an interesting but expensive potential weapon.

TNT also had one unusual property—it could be used, in small amounts, as a very effective catalyst in making clothing dyes. And far less TNT was needed for the clothing industry than for munitions, which meant that far less toluene had to be made. For the four decades after its invention, dye makers cooked up TNT for dyes that were used in making shirts, pants, dresses, and other types of clothing and fabrics. Amazingly, there was never a reported explosion involving TNT while it was used to make dyes.

A World War I munitions factory. Munitions factories began filling artillery shells with TNT, a powerful explosive that could be melted and poured with little risk.

In 1904, though, chemists found less expensive ways to distill toluene from oil and coal tars, which in turn made TNT an affordable alternative to other types of civilian and military explosives. It was the military, in fact, that was to get the most use out of TNT during the next century. Because TNT could be melted and poured without risking an explosion, munitions companies could fill artillery shells with larger, deadlier payloads than were possible using gunpowder or other types of explosives. Earlier exploding shells had been made using solid picric acid, a powerful explosive that had been discovered in the 1850s but had not seemed to offer the same commercial success as nitroglycerin had. It was not as safe to handle as TNT was, either, and was too unstable

SHAPED CHARGES

When a standard explosive charge detonates, the force of its detonation spreads out in all directions. In a borehole or other tightly confined space, almost none of the blast's energy goes to waste. In a less confined or open area, however, a lot of the energy moves away from the target rather than into it. This is where a shaped charge, which focuses the energy of its blast in one direction, can be useful.

A shaped charge is a charge of explosives with a cone-shaped or V-shaped hollow that directs the blast forward. The hollow inside the main charge is usually lined with glass, which melts in the heat of detonation to become a forceful projectile. The molten glass can punch through one foot of steel, or three feet of concrete, weakening the target for the explosion's shock wave.

Because the shaped charge can be easily attached to bridge supports and other standing targets with a claylike adhesive, it is often used in military demolition missions. Some military shaped charges have "standoff sleeves," scalloped-edged metal tubes at the base of the charge that allow for attachment to uneven surfaces. The force of shaped charge explosions can be increased by placing two such charges on opposite sides of a target such as a bridge support. When detonated at the same time, the shock waves from both charges collide inside the structure and reverse direction. Because the concrete of the bridge support cannot absorb the sudden change in the blasts' direction, the concrete crumbles.

Shaped charges have other uses than for military demolition missions. In steel plants, for example, shaped charges mounted on the end of long poles play a vital role near the end of the production process. Molten steel is poured out through vents in the bottom of some vats, vents that usually become plugged with cooled steel. The only way to unplug the vents is to blow out the mass of hardened steel. Using shaped charges, factory workers easily puncture and shatter these plugs.

when heated to be poured into bombs and artillery shells. And commercial explosives, like dynamite and blasting gelatine, were useless for military work. Although they could blow rocks apart easily enough, they did not have the power needed to destroy armored tanks and other battlefield vehicles.

What was needed, and what TNT offered, was a weapon that produced a huge shock wave that ruptured armor plating, destroying men and machinery alike. For a split second, the explosion of TNT created a force more than 120,000 to 150,000 times the pressure of the earth's atmosphere at sea level. In other words, each square inch of material against which a charge of TNT exploded—metal, wood, brick, or flesh—was subjected to the equivalent of between 1.8 and 2 million one-pound blocks being flung against it. This shattering force was called *brisance*, from the French word meaning "to break," and became a major measurement of a military explosive's effectiveness.

TNT and the First World War

The development of TNT as a useful tool of war came in time for it to be used in one of the bloodiest conflicts in history. On June 28, 1914, an assassin in the Bosnian town of Sarajevo killed Archduke Francis Ferdinand of Austria-Hungary, one of two Germanic empires in central Europe. The assassination of the archduke was an act of protest against the empire's invasion of Bosnia. It came at a time when the major powers of Europe—especially Germany and France—were looking for an excuse to go to war with each other. The murder of the archduke set off a chain reaction of troop mobilizations and declarations of war by the nations of Europe. Within a month, the great conflict that became known as World War I had begun.

With the outbreak of war, the need for explosives reached an all-time high. So much TNT was consumed during

TNT became an important military weapon. Its explosive power could rupture armored tanks, destroying the machinery and killing the men inside.

the war that some nations had trouble meeting the demands of their armed forces. Munitions companies had to come up with ways to stretch the supply of the explosive. The best way to do this, they discovered, was to mix TNT with other explosives that were less expensive but nearly as powerful. One of the first of these mixtures was Amatol, a combination of equal parts TNT and ammonium nitrate. TNT was used, alone or in combination with other explosives, in airplane bombs, artillery shells, antisubmarine depth charges fired from the decks of destroyers, and even in hand grenades. One of the most famous hand grenades—the German *stielhandgranate* (or "potato masher," as it was called by British and American troops)—consisted of a canister of almost pure TNT screwed onto a hollow wooden throwing handle.

A German pilot prepares to drop a bomb on Allied forces during World War I. TNT-filled airplane bombs and artillery shells turned much of Europe into a crater-strewn wasteland.

The use of TNT and other high explosives changed the way that war was carried out. By the middle of World War I, much of Europe—especially France—had been turned into a crater-strewn wasteland by airplane bombs and artillery shells. The troops on each side were forced to dig into the ground for protection, living in trenches along the edges of no-man's-land, which was dominated by machine-gun and artillery fire. In this war, death and destruction were brought about on a titanic scale. Some of the guns used in the war, including Germany's "Big Bertha" rail-mounted guns, fired shells that could make craters the size of two- and three-story houses. The sheer amount of explosives used was staggering. By the end of the war in 1918, millions of tons of TNT had been consumed by all the nations involved.

Shattering the Peace

War, with its titanic needs for munitions, had hastened the development of explosives as weapons of mass destruction. But with the end of World War I, most people figured that explosives would return to their prior role as tools of construction and mining. Most people hoped never to see war on a worldwide scale again, although they assumed that explosives had found a permanent place in the world's arsenals. What they did not expect was to find explosives used in peacetime against citizens and businesses, by political organizations waging their own private wars.

Germany's rail-mounted gun, "Big Bertha," fired shells that could make craters the size of large houses.

Though never intended for this purpose, explosives offered an easy tool for those who wished to further their political causes through violence. Gunpowder had been an effective tool in campaigns of terrorism. Anarchists—people who believed that all forms of government were evil and had to be overthrown by violent means—had used small, homemade gunpowder bombs to kill and injure people during the mid- and late 1800s. The invention of dynamite had given these terrorists an even more powerful tool for their war against society. And the explosives developed before and during World War I allowed even more violent and destructive acts.

One of the most devastating of these attacks took place in New York City on September 16, 1920. Shortly before noon, an unknown anarchist steered a horse-drawn wagon to a spot on the corner of Wall Street and Broad Street in New York City. He parked the wagon in front of the U.S. Assay Office, a government mining agency, near the New York Stock Exchange. The wagon was loaded with a few hundred pounds of TNT and small metal weights, such as those used to weigh down window sashes and the edges of curtains. One minute after noon, as many office clerks, stock exchange workers, and other people set out for lunch, the TNT-loaded wagon exploded.

The force of the blast, and the metal weights that rocketed in all directions, killed more than thirty people and seriously injured at least two hundred. Every window in the Assay Office, the Stock Exchange, and the nearby J. P. Morgan building shattered, and their walls were pitted by the weights that did not find human targets. (Some of these buildings still bear signs of the 1920 blast.) A reporter who rushed to the site soon after the blast described it as "such a scene as I had pictured as a possibility during the war [World War I], should the enemy succeed in dropping on the financial district one of his deadly aerial bombs."

The J. P. Morgan building after the 1920 anarchist attack. The TNT blast killed more than thirty people and shattered windows in nearby buildings.

Plastic Explosives

Senseless killings such as this one provided further evidence to many people that explosives were as powerful as they needed to be. In 1938, though, Europe was threatened by an old foe with a new face. Germany, which under the fascist dictatorship of Adolf Hitler had rebuilt itself from postwar ruin, was again trying to conquer the Western world. This time, however, it was being aided by Italy, and was assisting Japan in that nation's quest to conquer Asia and the Pacific. In World War II, the troops were not restricted to networks of trenches. Fighting took place along the roadways, in the cities, over the bridges, and throughout the countryside of Europe, and in the jungles of the Pacific. Each side realized they needed new types of weapons—including new types of explosives—to win this new world war.

Because troops were on the march far more than they were during World War I, there were many special requirements that these new explosives had to meet. They had to be stable enough to be carried for hundreds of miles without detonating. They had to be light enough for one soldier to carry. They had to be at least somewhat waterproof—World War II was a time of commando raids and underwater demolition teams, with specially trained soldiers and sailors conducting night operations to blow up bridges, ammunition depots, and other targets. Because these men would be swimming ashore for at least some of their missions, their explosives had to be made as resistant

The need for a water-resistant explosive arose during World War II when military operations often required soldiers to swim ashore for their missions.

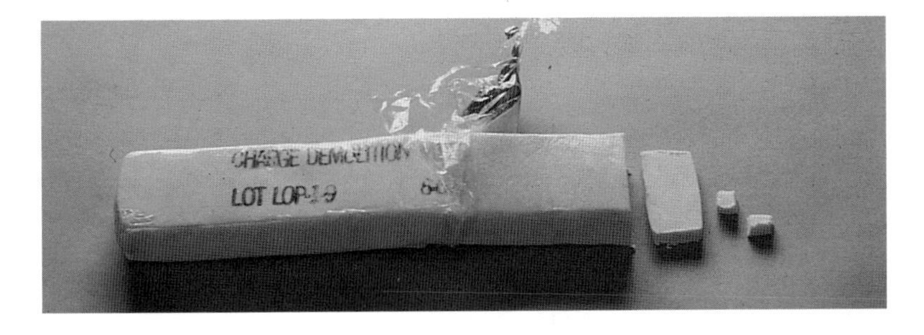

A bar of C-4 plastic explosive. Able to withstand temperatures ranging from –70 to 170 degrees Fahrenheit and extremely powerful, C-4 became a potent military and terrorist weapon.

to water as possible. And, ideally, the new explosives could be fastened to their targets without the hassle and time required for taping or tying them.

Two explosives were seen as promising candidates for the needed weapon: hexogen or RDX, a powerful, extremely sensitive explosive that was invented in 1899; and pentaerythritol tetranitrate, or PETN, an explosive as powerful as RDX and nitroglycerin but more sensitive than RDX, developed in 1901. These solid explosives were created in the attempt to make a more powerful substance than TNT or dynamite. Powerful they were, but they also were extremely sensitive, exploding at the mildest shock. They were too dangerous to use in large quantities, but in small amounts they could be handled safely, and were used as primers for TNT bombs and artillery shells.

Because explosives like RDX and PETN were so powerful already, explosives experts decided to use them rather than work to develop new explosives from scratch. The problem was to find a way to stabilize the pure explosive, to mix it with other materials that would make it transportable. Most of this work took place in Great Britain, which declared war on Germany in 1939, after watching Hitler's forces successfully invade the central European nations of Austria, Czechoslovakia, and Poland. Working rapidly, British chemists tried a number of mixtures before coming up with a combination of roughly 80 percent RDX and 20 percent wax, motor oil, and other materials that shielded the explosive from rough handling. These fillers, as they were called, did not contribute anything to the detonation of the RDX, but they did not absorb or weaken the effects of the blast either. And, most importantly, they made the RDX safe to handle and easy to mold to the side of almost any target. Because this solid explosive was so pliable, it became known as plastic explosive.

When the United States entered World War II in 1941, the U.S. Army began its own work with the British-developed plastic explosive. U.S. Army experts modified the original formula, increasing the amount of RDX to 88.3 percent and decreasing the fillers to 11.7 percent. Called Composition Four (or C-4), the army's version could be used virtually anywhere. It could stand temperatures ranging from –70 degrees Fahrenheit to 170 degrees Fahrenheit. And best of all, less than a pound of C-4 could seriously damage or destroy any bridge, building, or other structure.

Other Explosives of World War II

Mixing old explosives in new ways led to other powerful combinations besides

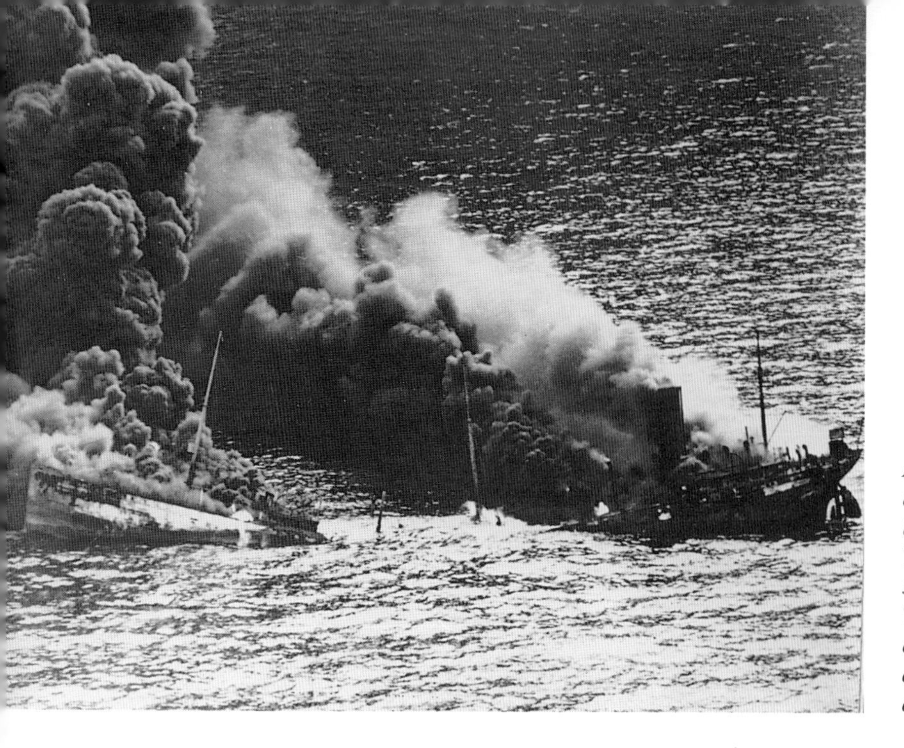

An Allied tanker sinks after being split in two by a German torpedo. Torpedoes require a strong mixture of RDX, TNT, and powdered aluminum to deliver an explosion forceful enough to sink a ship.

plastic explosives. Munitions makers began mixing RDX and PETN with TNT, which was still the king of military explosives. Combinations of TNT and RDX or PETN proved to be more powerful than TNT alone. And, much to the delight of many experts, they possessed many more times the shattering power of TNT. For example, an explosive called Cyclotol, which combined sixty parts RDX and forty parts TNT, produced a force of more than four million pounds per square inch from each explosion. In other words, for a split second, the shock wave and gases produced by Cyclotol exerted more than 270,000 times the pressure of earth's atmosphere at sea level.

At times, however, rather than shattering their targets, bombs merely needed to knock them down and, if possible, set them ablaze. For this type of mission, munitions companies found that they could get the desired effect by adding powdered aluminum to TNT, creating a mix called Tritonal. The burning aluminum gave off a great deal of heat, increasing the bomb's ability to start fires. At the same time, the aluminum absorbed part of the bomb's shattering power. The reduction of the shattering power served two functions: it kept the targets it hit from disintegrating, thereby ensuring that they would burn rather than break apart, and it prevented the shock of the blast from putting out the fires as they got started.

Torpedoes, too, required explosives that did not create sharp, shattering blasts. The goal of a torpedo is to detonate just below the center of a ship, creating a surge of power that lifts the ship out of the water and cracks its keel. With its keel broken, the ship splits in two and sinks. Submarine sailors called this technique "breaking the ship's back." But because part of the torpedo's explosion was absorbed by the water around the ship, a simple mixture of TNT and aluminum was found to be too weak. Instead, both the British and the American navies used a mixture of RDX, TNT, powdered aluminum, and

DETECTING PLASTIC EXPLOSIVES

Scientists designed the thermal neutron activation scanner to detect plastic explosives hidden in luggage.

Plastic explosives can be hidden almost anywhere and cannot be detected by standard X-ray machines or metal detectors. This is why plastic explosives are commonly used by terrorists. For example, the 1988 bombing of Pan Am Flight 103 over Lockerbie, Scotland, was caused by plastic explosives. To combat this threat, scientists are working on sophisticated devices for detecting plastic explosives.

One device, the thermal neutron activation scanner (pictured above), was designed to detect explosives hidden in luggage. A conveyor belt in the scanner takes luggage through a stream of neutrons. As the luggage and its contents absorb the neutrons, their atoms give off brief bursts of a type of energy called gamma rays. A sensor in the scanner analyzes the pattern and strength of the gamma rays to determine if plastic explosives are present.

The same process cannot be used to find plastic explosives carried by passengers, as neutron bombardment would likely kill any person who walked through the stream. But another type of detector, a vapor detector, can be used to seek out bombs carried by passengers. Vapor detectors test the air around passengers for fumes given off by plastic explosives. This type of detector can find one or two molecules of explosive vapor in a sample of one trillion molecules of air.

These devices, though still experimental, offer much hope for the future. Presently, however, thoroughly searching passengers and their luggage by hand has proven itself to be the most successful method for finding bombs. El Al, Israel's state-run airline, is one of the few airlines that conducts such searches. In 1986, a search of this type turned up a passenger trying to carry a two and one-half pound block of Semtex aboard a flight to London. Terrorism experts not connected with the airline think that vegetable oil leaked from the Semtex in the warm April weather, attracting the attention of El Al inspectors. El Al has never revealed all the details of the case.

(in the U.S. Navy) a desensitizing agent to let the torpedo withstand shock and rough handling without blowing up.

Postwar Cutbacks

With the end of World War II in 1945 came a large round of military reductions and a new international commitment to world peace. Development of most new weapons, including explosives, was halted. The only major weapon being tested was the atomic bomb, a new type of explosive that broke atomic, rather than molecular, bonds to obtain its power. It was hoped that the end of the war would mean the end of the type of destruction that the world's new explosives had brought about. And it was hoped that the end of war would mean that civilians would not have to worry any more about being victims of violence. But just as anarchists had found ways to use TNT for their campaigns of terror, modern terrorists discovered ways to use the explosives of World War II for their political needs.

Ironically, the same plastic explosives developed to free the world from the terror of Nazi fascism became one of the major weapons of the twentieth-century terrorist. The same properties that made plastic explosives ideal for many military missions also made them ideal for terrorist campaigns. They were both light and powerful. Five pounds of C-4, for example, could cause a more violent explosion than did the wagonload of TNT in the 1920 Wall Street bombing. And most acts of terror required far smaller amounts of explosives than that.

Plastic explosives made their debut as terrorist weapons in the early 1960s. At the time, the nation of Algeria was seeking independence from 130 years of French rule. Many people in France

Post-World War II explosives research concentrated on the development of the atomic bomb, a new type of explosive that obtained its tremendous power by breaking atomic, rather than molecular, bonds.

Algiers during the 1960 revolt. Members of a group called the Secret Army Organization detonated C-4 explosives around Algiers after failing in their attempt to topple Algeria's regional government.

opposed Algerian independence. A group of French army officers, which called itself the Secret Army Organization, sided with this anti-independence sentiment and tried to take over Algeria's regional government. When that attempt failed, the group turned to terrorism. Its members stole some C-4 explosives from their units' armories and used the explosive in bombs that they detonated around Algiers. C-4 also was used in France in a failed attempt to assassinate the French president, Charles de Gaulle.

C-4 was, and still is, a popular choice among terrorists, but it was joined by a number of other powerful alternatives in the late 1960s and the 1970s. Chief among these other explosives is Semtex, a compound of mostly RDX and PETN, with a small amount of fillers and binders (mostly vegetable oil). Semtex, a product of the former communist nation of Czechoslovakia, was used to blow up Pan Am Flight 103 over Lockerbie, Scotland, in 1988. Semtex was developed in the early 1970s for use by terrorists who were friendly to Czechoslovakia and, more importantly, to the former Union of Soviet Socialist Republics (USSR). Its main virtue as a terrorist weapon was that a small amount could do a great deal of damage. Whereas a pound of dynamite could destroy a midsize car, for example, a pound of Semtex could easily destroy

Police officers remove a body from the wreckage of Pan Am Flight 103, which was blown apart by a terrorist's bomb over Lockerbie, Scotland, in 1988. Investigators believe the bombing was done with no more than half a pound of the powerful explosive Semtex.

most of a one-story house. Usually, far less than a pound is needed to have a devastating effect on a target. Investigators into the Lockerbie bombing have speculated that the airliner was knocked out of the sky by no more than half a pound of Semtex—small enough to fit, with a timer, in a cassette player.

The USSR actually controlled the distribution of the deadly explosive, sending it to groups seeking to establish communist governments in noncommunist nations and to terrorists who had declared war on nations hostile to the USSR. So much Semtex was produced and distributed that, according to Vaclav Havel—Czechoslovakia's first president after the collapse of the communist government—"world terrorism has enough Semtex to last 150 years."

Fortunately for the world, not all of the advances in explosive technology have been made for military and terrorist use. At the same time that plastic explosives were being adapted for use as terrorist weapons, the commercial blasting industry was going through a revolution of its own.

CHAPTER 5

The Commercial Explosives Revolution

Most of the explosives made during the twentieth century have been made for military use. In the first half of the twentieth century, America alone produced more than thirty-two million tons of explosives, mainly for use in the two world wars. Although the most devastating explosives—from TNT to RDX-based plastic explosives—were designed for war, the most widely used and versatile explosives have been designed for commercial use.

Until the mid-1950s, virtually all commercial blasting was done with dynamite. Dynamite was stable, reliable, and powerful, three virtues that appealed strongly to professional blasters. One of the drawbacks of dynamite, its high freezing point, was solved by the 1920s. Explosives chemists discovered that adding ethylene glycol dinitrate, a chemical similar to nitroglycerin, to dynamite lowered the freezing temperature of the entire mixture. The added chemical acted much like antifreeze, which lowers the freezing temperature of water in car radiators. With this addition, miners and blasters could use dynamite in virtually any weather without having to warm it up, a process that often caused unexpected explosions.

Even so, dynamite posed some difficult, and sometimes deadly, problems. Because it contained nitroglycerin, it still caused serious headaches in those who used it. Dynamite that had been left sitting around for a while was an extremely risky substance. The nitroglycerin tended to leak out of its surrounding dope and through the cardboard tube and waxed paper that held it together—a process called sweating. When that happened, any serious shock, from dropping it on a hard floor to accidentally hitting it with a hammer, could set it off.

Ammonium Nitrate Returns

Dynamite's flaws led many researchers to continue the search for an explosive that was easier and safer to use. Just as nitric acid had offered nineteenth-century scientists a good starting point for the creation of high explosives, ammonium nitrate offered a good starting point for the creation of safer high explosives. The Swedish scientists Norrbein and Ohlsson had proven that ammonium nitrate, although only 70 percent as powerful as dynamite, could be a useful explosive. It had only two drawbacks. It could only be detonated by a powerful primary explosive, which meant buying and using dynamite or TNT, the two explosives researchers were trying to replace. Also, ammonium nitrate quickly absorbed any water with which it came in contact. In mining or construction blasts, explosives were placed in deep holes, called boreholes, that were drilled into the blasting area. Any water—from morning dew, rainwater, groundwater seepage,

Until the mid-1950s, most commercial blasting was done with dynamite. (Left) Workers drill deep holes, called boreholes, into the blasting area. (Below) The boreholes are then filled with dynamite, which is wired to a detonator.

or any other source—could interfere with a blast.

At first, overcoming these two problems seemed simply a matter of sealing off the explosive in some kind of water-repellant material without lessening the force of the blast. In 1935, researchers at the American explosives firm E.I. du Pont de Nemours came up with a simple solution. They mixed ammonium nitrate with paraffin (candle wax), both to stabilize the explosive and to add heat to the detonation, and sealed the mixture in metal cans. The top of each can could be screwed into the bottom of another. The primer, a charge of TNT, went in a separate can that also could be attached to the cans of ammonium nitrate and paraffin. Together, they were sold as a single explosive product, which Du Pont called Nitramon.

BLASTING CAP

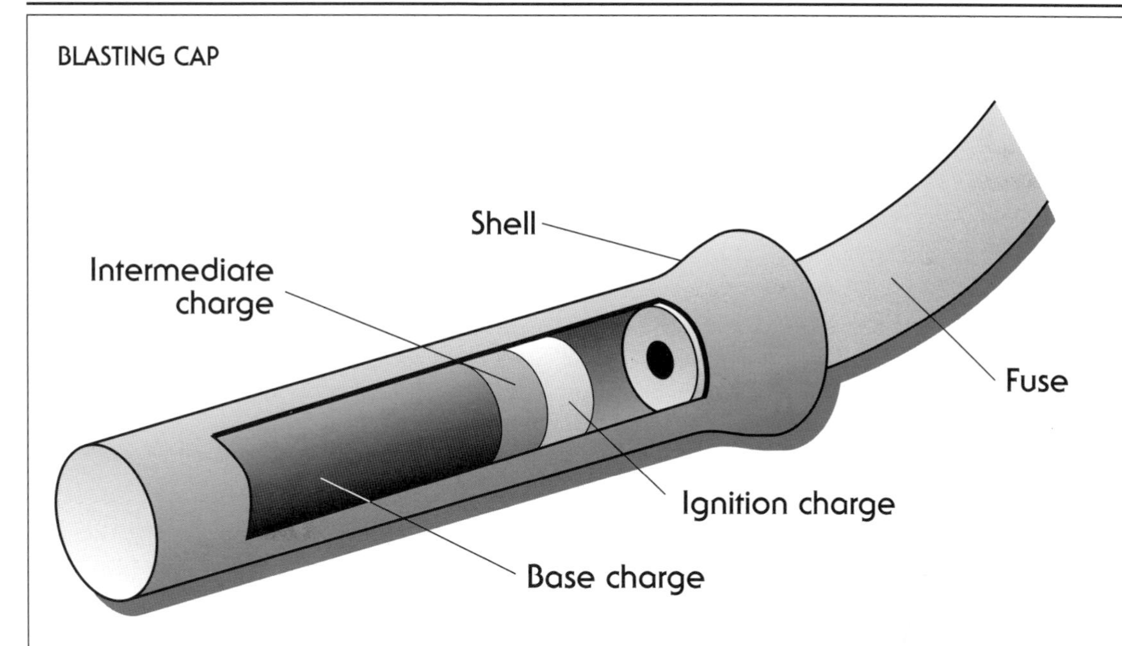

A blasting cap is a small metal cartridge filled with a small charge of gunpowder or high explosive. Blasting caps are used to safely detonate powerful explosive charges such as dynamite or plastic explosives.

A blasting cap's explosion actually is a chain reaction of detonations in which small charges ignite progressively larger ones. The first link in this chain is the ignition charge, a tiny amount of extremely sensitive explosive. The blast of the ignition charge ignites a less sensitive but more powerful intermediate charge. The intermediate charge's detonation in turn ignites an even more powerful base charge. The explosion of the base charge causes the main charge to detonate.

One type of blasting cap detonates with the aid of a detonating cord, or fuse. This cord contains a core of black powder or high explosive such as PETN. Lighting the fuse transfers energy to the blasting cap's ignition charge. Another type of blasting cap is triggered by an electric pulse sent through a fine wire inside the cap's metal shell. This wire, called a bridge wire, takes the place of the fuse and sets off the ignition charge. It also allows for more precise control over the timing of a series of blasts. However, this method cannot be used if there is a risk that lightning or static electricity might set off the explosion prematurely. In these cases, detonating cords are used instead.

Both types of blasting caps must be handled with extreme care. Simply dropping a blasting cap is enough to trigger the chain reaction inside the metal shell. While blasting caps are less powerful than the main charges they are designed to detonate, they are powerful enough to maim, blind, or kill anyone who mishandles them.

Nitramon offered the power of ammonium nitrate without the risk of water contamination. Because of Nitramon's unique connectable-can packaging, in which each can held specific amounts of the ammonium nitrate mixture, blasters could easily measure the right amount needed for a blast. They simply had to connect enough cans of the ammonium nitrate mixture and screw on a can of the TNT primer. The cans kept the two explosives safe from any water that might be in the boreholes, allowing for a reliable detonation.

Nitramon was far safer than dynamite. The TNT primer and the main charge of ammonium nitrate could only explode under certain conditions. The primer would not detonate unless it was set off by a blasting cap, and the main charge would not detonate until it was set off by the primer. Thus, Nitramon was safer to ship, store, and use than was dynamite. Nitramon's low price made it a nearly ideal explosive for mining, tunneling, and other types of large-scale construction blasting. Du Pont officials called their new product a "blasting agent" to distinguish it from other types of explosives, such as dynamite and TNT.

As safe and economical as it was, however, Nitramon was not quite the wonder-explosive professional blasters were looking for. Unless the cans containing the Nitramon fit exactly into the borehole, the full effect of the explosion was lost. Any gaps between the can and the wall of the borehole dispersed some of the force of the blast, breaking rock unevenly and sometimes requiring a second shot to finish the job. This was not desirable, because additional blasts both increased the danger of injury and took up valuable construction time. Clearly, the idea behind Nitramon could be improved. There had to be some way to mix or package ammonium nitrate-based explosives so they would fill boreholes completely, yet still resist water. But until shortly after World War II, no one knew how to do that.

Fire in the Hold

In 1947, two horrific explosions involving ammonium nitrate pointed the way toward an explosive that would solve these problems. One of these explosions took place on April 16 in Texas City, a seaport northwest of Galveston, Texas; the other took place a few months later in Brest, a seaport on the northwest coast of France. Both explosions involved ships carrying loads of ammonium nitrate fertilizer.

A Du Pont scientist pounds on Nitramon, demonstrating why the new blasting agent is called "the safety explosive." Because it can only be detonated by a blasting cap, Nitramon is safer to store, ship, and use than dynamite.

(Left) The explosion of 54,463 pounds of Nitramon at a Pennsylvania rock quarry. (Right) Tons of shattered rock lie at the foot of the quarry after the blast.

In the Texas City explosion, a ship named the *Grand Camp* was being loaded with the fertilizer when a small amount somehow caught fire. The fire filled one of the ship's holds with smoke. Attempting to put out the fire, the crew shut the hatches and opened a valve that let steam into the hold. They hoped that the fire would burn up all the oxygen in the hold, and that the steam would help smother the flames. Instead, the fire worsened; the burning ammonium nitrate released enough oxygen to keep the fire going, and the steam actually acted as a blanket, holding in the heat from the fire.

Within an hour, the hold of the *Grand Camp* became so hot that the remaining fertilizer exploded, erupting into gas and flames in less than a second. The explosion was as bad as if the hold had been filled with pure nitroglycerin. It destroyed the ship, killed or injured more than thirty-five hundred people, set off explosions at a nearby chemical plant, and sent out earthquake-like shock waves that could be felt by people in a four hundred-mile-wide circle around the ship. The shock and pressure created by the explosion were so powerful that two airplanes flying nearby were knocked out of the air, spinning out of control until they hit the ground. The explosion caused more than $50 million worth of damage.

The Brest explosion also caused widespread damage. It killed twenty-one people and injured one hundred others, but the force of the blast was the biggest surprise. The explosion of the 6.6 million pounds of ammonium nitrate in the Brest harbor caused greater damage to buildings and spread over a wider area than did the explosion in Texas City. Some dockside buildings near the ship were obliterated, and windows were cracked or broken as far as ten miles from the blast.

The Texas City Grand Camp *explosion resulted when a small amount of ammonium nitrate fertilizer caught fire. The powerful explosion killed or injured more than thirty-five hundred people, and caused more than $50 million worth of damage.*

The news of the titanic amount of damage caused by both loads of exploding fertilizer shocked the world. It seemed inconceivable that a shipload of fertilizer could explode at all, much less with such great force. The fact that it had happened, though, amazed chemists who were seeking a more effective ammonium nitrate explosive. Until then, it had just been assumed that ammonium nitrate was too chemically stable to detonate without a primer of high explosive. The Texas City and Brest explosions, however, had involved ammonium nitrate that somehow had caught fire, had baked for a while, and then had flared to detonation. Chemists in the explosives industry realized they would be able to create a stable yet powerful explosive if they could reproduce similar explosions under controlled conditions. Work began immediately.

ANFO

Experiments proved that carbon from the paraffin-covered paper bags that contained the fertilizer had given the explosions their devastating intensity. This led scientists to conclude that the key to inventing a more powerful explosive lay in some combination of carbon and ammonium nitrate.

Scientists tried a wide range of carbon-bearing materials in their quest for the new explosive. Coal, motor oil, and mixtures of coal dust and oil all worked, but they worked poorly. They created weak explosions and gave off a lot of sooty, black smoke. Finally, in 1955, two coal company chemists, Hugh B. Lee and Robert L. Akre, came up with a mixture of ammonium nitrate pellets, or prills, and carbon black, a type of carbon-rich soot. As long as it stayed dry, this mixture, which they called Akremite, created a powerful explosion. Like other ammonium nitrate explosives, however, it soaked up water like a sponge, forming a pastelike mess that could not be detonated. To protect it from groundwater or other moisture, Lee and Akre suggested wrapping Akremite in plastic. Boreholes also had to be lined with plastic to protect the water-loving mixture.

To some skeptics, Akremite seemed just too good to be true. Not only was it easily made, and effective as long as it stayed dry, but it cost almost nothing.

These characteristics won over many people in the mining and blasting fields. Joe Dannenberg, an explosives historian, summed up the main reason that Akremite promised to revolutionize commercial blasting: "All you had to do was rig up a barrel mixer [such as that used to mix cement], and get yourself some polyethylene bags and, presto, you were in the explosives business at your own operation." Many people believed that these and other benefits of the ammonium nitrate explosive outweighed its drawbacks and the doubts about its usefulness. Soon, Akremite was being used in a few mining and commercial blasting operations, mostly for coal mining.

However, other chemists and professional blasters continued to experiment with different chemicals. Within two years of the introduction of Akremite, a number of blasting companies began using a more efficient mixture, one that substituted liquid fuel oil for the dry carbon black. As the fuel oil mixed with the ammonium nitrate prills, it created a slushy mixture that could be poured right into the borehole. Like the Sprengel explosives that were popular in Europe, the ingredients of the new explosive could be mixed at the blast site. There was little or no danger in transporting them. And best of all, the most expensive ingredient was the oil, which at the time was not very expensive. This combination of ammonium nitrate and fuel oil—called ANFO—forever changed commercial blasting, allowing bigger blasts at a lower cost. Within ten years, ANFO dominated the civilian explosives industry in America, making up more than 60 percent of the total amount of explosives used by blasters.

The Need for Water-Resistant Explosives

ANFO was fine for blasts in dry areas where it needed little, if any, protection from moisture. It was ideally suited for open-pit mines in deserts and open plains. But it did not perform terribly

An explosion at an open-pit iron mine. ANFO, the leader in the civilian explosives industry, performed well in dry areas like open-pit mines. However, its poor performance when wet led chemists to search for a more water-resistant commercial explosive.

well in areas where it might get wet. Even when coated with fuel oil, the prills absorbed water, deadening or eliminating ANFO's explosive properties. For these more humid climates, blasters still needed something that would work even when it got wet. Fortunately, such an explosive was just about ready for market.

Du Pont, which was and is one of the world's leading explosives manufacturers, had decided to tackle the problem of water resistance head-on in the 1940s. Some of the company's chemists worked to improve Nitramon, making it powerful enough to do its work with only one blast. Others sought a way to make the ammonium nitrate overcome the dampening effect of water without using special packaging. These chemists knew that water did not destroy the explosive power of ammonium nitrate. All the water did was act as a shock absorber, suppressing the explosive's reaction to large shocks or flame. Wet ammonium nitrate that was allowed to dry out was just as explosive as ever.

Tovex

This insight into the relationship between water and ammonium nitrate led the Du Pont chemists to the following line of reasoning: The problem with wet ammonium nitrate was not that it could not explode. It simply was too stable when wet to respond to the shock created by a primary explosive. The answer to the problem, the chemists decided, was to find some way to make ammonium nitrate explode while it was mixed *with* water. An explosive mixture of water and ammonium nitrate also would give the company an easily poured explosive that would fill boreholes completely.

From 1942 through the mid-1950s, Du Pont chemists experimented with a number of sensitizers, chemicals that made the ammonium nitrate sensitive to slightly less forceful shock waves. They combined this mixture of ammonium nitrate and sensitizers with water, adding thickeners that turned the watery compound into a syrupy gel. Turning the compound into a gel made it easier and safer to transport the explosive to the blast site and to pour it into its boreholes. As a gel, the compound was able to disperse the effects of most heavy blows and shocks. The first of these water gel explosives, as Du Pont named them, were perfected by 1950, but were too expensive to compete with other types of explosives. In the mid-

Du Pont chemists worked to improve Nitramon through the 1940s and early 1950s. They hoped to increase its water resistance and simplify its use.

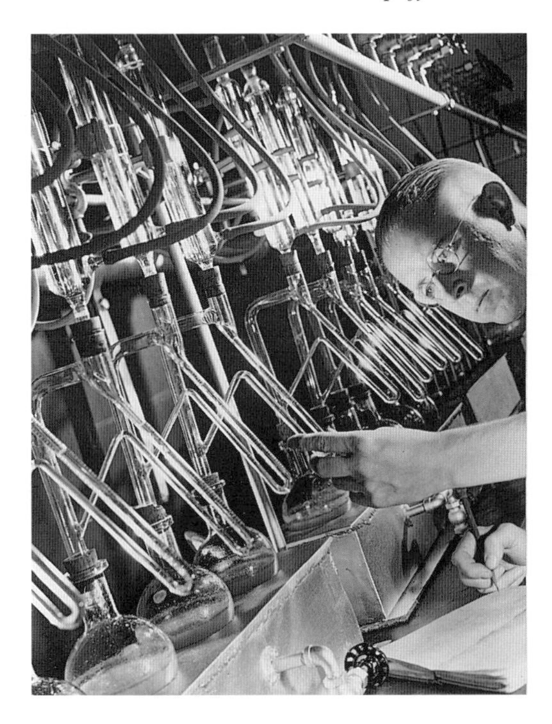

New York Coliseum construction workers load charges of Du Pont's water gel explosive, Tovex. The explosive's gelatin form made it safer to transport and easier to pour into boreholes than other explosives.

1950s, however, the cost of making water gels had dropped, and the company began selling its product under the name Tovex.

At first, Tovex could only be detonated in large boreholes using large, high explosive primers. Even though the sensitizers in the mix allowed the wet ammonium nitrate to explode, it could only do so in large quantities. Small amounts of Tovex exploded weakly, if at all. The need for large amounts of the explosive restricted its use to very large-scale operations, such as the huge iron mines in the Mesabi Range area of Minnesota. Over the next decade, however, the company developed improved forms of Tovex that could be set off using ordinary blasting caps.

Tovex sold well both before and after these improvements. In its first three years on the market, more than forty-five million pounds of Tovex were sold in the United States. By 1974, mines and other blasting operations

Workers load gel explosive into boreholes. The explosives will be detonated by delay-action electric blasting caps that fire the holes in sequence.

BUILDING DEMOLITION

Building demolition requires careful planning and precise work. A typical demolition job starts weeks in advance of the demolition date, as workers take out anything that might prevent the building from coming down smoothly. They hammer through the lower floors and tear out walls, doors, stairways, and other structures that do not hold up the building.

A few days before the blast, engineers place charges of high explosives throughout the lower floors of the building. A typical fifteen-story building needs about four hundred explosive charges in order to safely topple. Most of the high explosives are placed in the support columns on the bottom floor because those columns bear the most weight. Every charge is paired with a blasting cap that will ignite the explosive. Every blasting cap is connected with electric wire to a blasting box located outside of the building.

When everything is in place, a technician detonates the explosives by sending six hundred volts of electricity to the blasting caps. The explosives ignite in a sequence, each a half second apart, beginning at the front of the building and moving quickly to the rear. As the front columns collapse, the building begins to lean. When the next columns go, the first floor caves in, dropping the full weight of the building through the bottom floors. The force of the drop snaps columns on the middle floors and then throughout the building. At this point, the structure is so weak that the building tears itself down. When the blast is done correctly, the building appears to collapse in on itself and leaves neighboring buildings untouched.

were consuming nearly three hundred million pounds of the Du Pont water gel and of similar explosives made by other companies. Tovex was so successful that, in 1974, Du Pont announced it would stop making dynamite by the end of 1976 and concentrate on Tovex.

This news shocked many people in the explosives, mining, and construction industries who had already seen ANFO take over much of the work that dynamite had done. Du Pont's action did not mean that dynamite was obsolete, and many other explosives makers kept their dynamite factories running. But the use of dynamite for blasting has fallen off since 1974. It now accounts for less than 4 percent of all explosives sold. Dynamite still is needed for a number of tasks: to break up coal in coal mines, a task accomplished by types of dynamite called permissibles, which do not ignite natural gas or coal dust; to blast apart rock that is very hard; and to detonate more powerful explosives that can only be detonated by an explosive as powerful as dynamite. It also is one of the best commercial explosives for use in very cold weather.

Other Commercial Explosives

Civilian use of explosives these days does not stop with mining or construction. Tiny explosive devices are being used for an increasing number of everyday applications, many of which actually help save human lives. One of the most common of these modern explosive devices can be found in automobile air bags. When a car is involved in an accident, usually a head-on collision, a sensor triggers a tiny explosive charge of sodium azide. The explosion creates a huge volume of nitrogen gas, which inflates the air bag in roughly one twenty-fifth of a second—literally in less than the blink of an eye.

Dynamite is still used for some blasting jobs. Here, miners use a machine to drill holes in the rock face in preparation for a series of dynamite blasts.

Small explosive charges perform many critical jobs aboard the U.S. space shuttles as well. The bolts that connect the shuttle to its solid rocket boosters and liquid fuel tanks contain tiny amounts of high explosives. During launch, these explosive bolts are detonated to separate the boosters and tanks from the shuttle. Eight explosive bolts also hold the entire shuttle assembly to its launch platform before and during launch, exploding and vaporizing after the engines have built up enough thrust to lift the vehicle into the air. Small explosive charges help launch satellites from the shuttles' cargo bays into orbit, and also open antennas and solar energy panels on the satellites themselves. Explosives even help the shuttles return from orbit by forcing their heavy landing gear down.

By far the most dramatic civilian use of explosives is to demolish large buildings, especially those in the middle of cities. Building demolition is not so much a matter of blowing up buildings as it is getting them to topple in on themselves. Demolition crews actually use fairly small amounts of explosives (usually dynamite)—only enough to shatter supporting walls and pillars in strictly defined patterns that collapse the buildings in a particular direction. One of the most recent and most spectacular demolition jobs was the 1993 destruction of the thirty-eight-year-old Dunes Hotel in Las Vegas. Formerly one of the jewels of the Las Vegas Strip, the Dunes had been losing business to a

Small explosive charges perform vital functions aboard U.S. space shuttles, including helping to launch the shuttles and their satellites into orbit.

Explosive charges inflate an automobile air bag in less than one second. The bag inflates when a sensor triggers a tiny explosive charge of sodium azide. The explosion creates nitrogen gas which inflates the air bag.

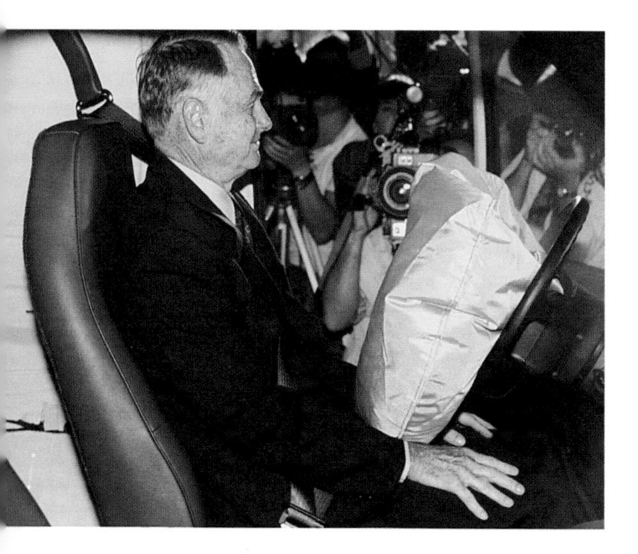

Building demolition requires weeks of careful planning to ensure proper placement of explosive charges. When done correctly, the building appears to collapse in on itself, leaving nearby buildings untouched and a pile of rubble where a high-rise building once stood.

series of huge casino-resort hotels that had been constructed during the 1980s and early 1990s. Its owner, casino magnate Steve Wynn, decided to demolish the building to make way for a more modern establishment. Thus, on October 27, 1993, 365 pounds of dynamite and 450 gallons of highly flammable aviation fuel were detonated throughout the twenty-three-story hotel. Reporters and other spectators said that the building seemed to hover in midair before collapsing in on itself. When the show was all over, all that was left was a thirty-foot-high pile of rubble that was still smoldering a day later.

High explosives have played a huge role in the development of the modern world simply through their power and versatility. Explosives help nations pull raw material for roads, buildings, and other structures out of the ground. They help workers clear and shape land as these structures are constructed. And they help dismantle these structures when they no longer are needed. Dynamite helped spur the development of the modern world, allowing people to more efficiently shape and control their nations. In much the same way, modern high explosives are one of the forces that have kept this development going.

CHAPTER 6

The Future of Explosives

The chemical explosives that exist today are probably as powerful as they are ever going to get. For one thing, it is highly unlikely that any company will want to take the risk of developing and making more powerful explosives. As the power of an explosive increases, the dangers involved in making it increase as well. But even if these dangers could be easily overcome, there still would be no real need to create more powerful explosives. The explosives that are available today offer as much power and do as much work as any group, civilian or military, will ever require.

There are two current trends in explosives research, however, that have nothing to do with making stronger explosives. One of these trends involves seeking out ways to make the explosives already in production still safer to use. The other is finding ways to use explosives for tasks other than simply breaking apart rock and moving earth. In the future, explosives may be used to make the materials that will advance humanity's ability to use information.

Improving Explosive Safety

One of the chief centers for safety-related explosives research is the Energetic Materials Research and Testing Center near Socorro, New Mexico. Run by the New Mexico Institute of Mining and Technology, the center studies ways to improve explosives and to apply them to different tasks. Scientists at the center have found ways to stabilize explosives that are far more advanced than any method ever tried. As solid explosives are formed, their molecules gather together in crystals. The method developed at the center forces a plastic compound into this pattern of crystals, essentially wrapping each crystal in a protective shell. The plastic protects the explosive not only from shock but from heat and from chemicals that could either destabilize or weaken the explosive. This added isolation ensures that the explosive will only go off when triggered by the proper detonator.

One industry that already has a use for these plastic-bonded explosives is the oil industry. Oil companies use explosives to clear out sand and other debris that block oil flow in deep wells. Some of these oil wells go down as far as three miles. At those depths the heat from inside the earth can reach temperatures of hundreds of degrees Fahrenheit. The wells become so hot that the explosives can and do detonate prematurely, wrecking the walls of the wells and sending more debris to the bottom. Plastic bonding promises to give the explosive charges a better chance of reaching the bottom of the wells intact.

Researchers are looking for other ways to give blasters greater control over how and when their charges explode. Most construction and mining blasts use a series of explosives ar-

Explosives clear out sand and debris that block oil flow in deep ocean wells. The development of plastic-bonded explosives holds great promise for the oil industry because explosives can detonate prematurely at such depths.

ranged to explode in a pattern. These patterns are designed to force the explosion to move rock or earth in a specific direction. Also, each charge or group of charges is set off in a staggered order, again driving the force of the explosion in a specific direction. The more control a blaster has over when each charge explodes, the safer and more effective the overall blast.

Alfred Nobel's original mercury fulminate blasting caps have given way to a number of different detonating methods. Some modern blasting caps use small amounts of PETN or RDX that are set off by less stable explosive chemicals (which themselves are triggered by a modern version of William Bickford's tightly packed gunpowder safety fuse). Others use an electric spark, usually generated by a battery pack, to trigger the primer charge. Electric firing allows more control over exactly when each charge explodes. In some cases, the delay between two sets of charges can be as long as twelve seconds or as brief as twenty-five milliseconds (one-fortieth of a second). Primer charges also may be set off by blasting cord, a thin tube of PETN or RDX that is attached to a chemical or electric blasting cap.

Most modern commercial blasting jobs are done using some sort of electric firing device. But using electrical triggers carries its own risk. A stray electric current or a static charge can set off the blast before the blasters are ready. For this reason, professional blasters will not even set up a shot if there is a risk of lightning striking nearby. One way to avoid the risk of an accidental explosion would be to develop a computerized detonator. Such a device would be able to tell the difference between an intentional firing signal and a static charge or, possibly, a power surge on the line. This ability would allow blasters both a safer system for detonating explosives and a more precise way to control the pattern in which they explode.

One possible component of this computer-controlled detonator has already been developed. In 1987, Sandia National Laboratories and the University of New Mexico, Albuquerque, developed a printed circuit that can act as a fuse, replacing many types of wire-based electric fuses currently used. Smaller than one-sixteenth of an inch square, the device consists of an H-

A worker prepares a dynamite blast to be detonated by an electric firing device. Much modern commercial blasting is done with electric triggers, which can be set off accidentally by static electricity or other stray electric sparks. The development of a computerized detonator could help eliminate this risk.

shaped copper form printed on a silicon chip. The crossbar of the H is heavily coated with phosphorus, a chemical that flashes into a very hot gas when heated. The inventors of the device, which is called a semiconductor bridge, say it could detonate explosives roughly one thousand times as fast as other electrical triggers. Although semiconductor bridges are still experimental, combined with a computerized detonator they could produce a finer edge of control over commercial blasts.

Protecting the Environment

Making explosives safer also means protecting the environment from their toxic effects. Most explosives, and the chemicals used to make them, are poisonous to most plants, animals, and people. Researchers all the way back to Ascanio Sobrero have known about the toxic effects of the chemicals used in explosives. When Sobrero was working with nitroglycerin in the late 1840s, he fed a small amount of it to a dog. Though the animal seemed to enjoy the sweet-tasting liquid, Sobrero reported, it died within a few hours. Sobrero knew that nitroglycerin tasted sweet because he tried a few drops of it himself, and almost immediately collapsed with fatigue.

More powerful explosives like RDX and TNT are even more potent poisons. Around some old explosives factories, they have soaked into the soil, poisoning local plants and animals. The bombs dropped on France and Germany during the two world wars have

The experimental semiconductor bridge, a printed circuit smaller than one-sixteenth of an inch, can detonate explosives one thousand times faster than other electric triggers.

Researchers at Sandia National Laboratories in New Mexico (pictured) are among those searching for ways to improve the safety of explosives.

created similar problems. Even today, these nations are laced with explosives that failed to detonate on impact, but buried themselves deep underground. These unexploded bombs pose the obvious risk of injuring or killing anyone who accidentally uncovers them. But they also pose the risk of soil or groundwater contamination if their loads of TNT, RDX, and PETN seep out of their rusting casings.

RDX also poses another problem. Ever since the beginning of the cold war, RDX was used as the primary explosive in nuclear weapons. In each nuclear warhead, the nuclear material (usually plutonium) is surrounded by a jacket of RDX. This high-explosive charge is designed to violently squeeze the plutonium, forcing it into the chain reaction that creates a nuclear explosion. Because the RDX is so close to the plutonium, over the years it becomes radioactive as well. With recent drives to dismantle the world's nuclear arsenals, there has been a scramble to come up with ways to dispose of the now-radioactive RDX jackets other than blowing them up and possibly spreading radioactive debris.

Recently, researchers in New Mexico, at Oak Ridge National Laboratory in Tennessee, and at the University of California, Los Angeles, discovered that some strains of bacteria in ponds and in soil may feed on TNT and RDX. Scientists studying contaminated soil and

A sign outside Oak Ridge National Laboratory warns of water contaminated by nuclear waste. Oak Ridge researchers have located strains of bacteria that may one day be used for cleaning contaminated soil and ponds near old explosives factories.

New York's bomb squad makes preparations for removal of a World War II aerial bomb found by workmen in 1969. Unexploded bombs usually contain live explosives that can detonate and kill or injure anyone who accidentally uncovers them.

ponds near old explosives factories noticed that large amounts of both types of explosives seemed to disappear. Further study showed that some types of bacteria in the soil and the water had become immune to the toxic chemicals and had actually learned to thrive on them. Currently, researchers are studying ways to isolate these bacteria and package them for use in areas that have been contaminated by explosives. The hope is that the bacteria will not only clean up explosive residue, but also will be able to feed on other toxic chemicals that are polluting the ground in other areas.

Widening Explosives' Commercial Uses

Although many of the explosives used around the world have been used to mine ores and dig out tunnels and roadways, a smaller but significant number are being used in an industrial process called explosive welding or cladding. Explosive welding is used to force two or more sheets of metal together to form a very strong, very tight bond. It is mainly used to join two different types of metal that cannot be successfully welded by conventional welding methods. For example, when the United States eliminated silver from its coinage in the 1960s, it replaced it with a more durable metal "sandwich" of copper and nickel. To ensure that the two metals stayed together, it used sheets of copper and nickel that had been blasted together.

The basic technique of explosive welding involves placing two sheets of metal together with an explosive above the top sheet. The explosion drives the upper sheet into the lower one at a slight angle. The force of the explosion drives the sheets together so fast that their facing sides momentarily melt and swirl together. At the same time, the very topmost layers of molecules spray out from between the sheets, forcing out any dust or dirt particles that might create gaps between them. Once they cool down, they are locked together in

EXPLOSIVE CLADDING

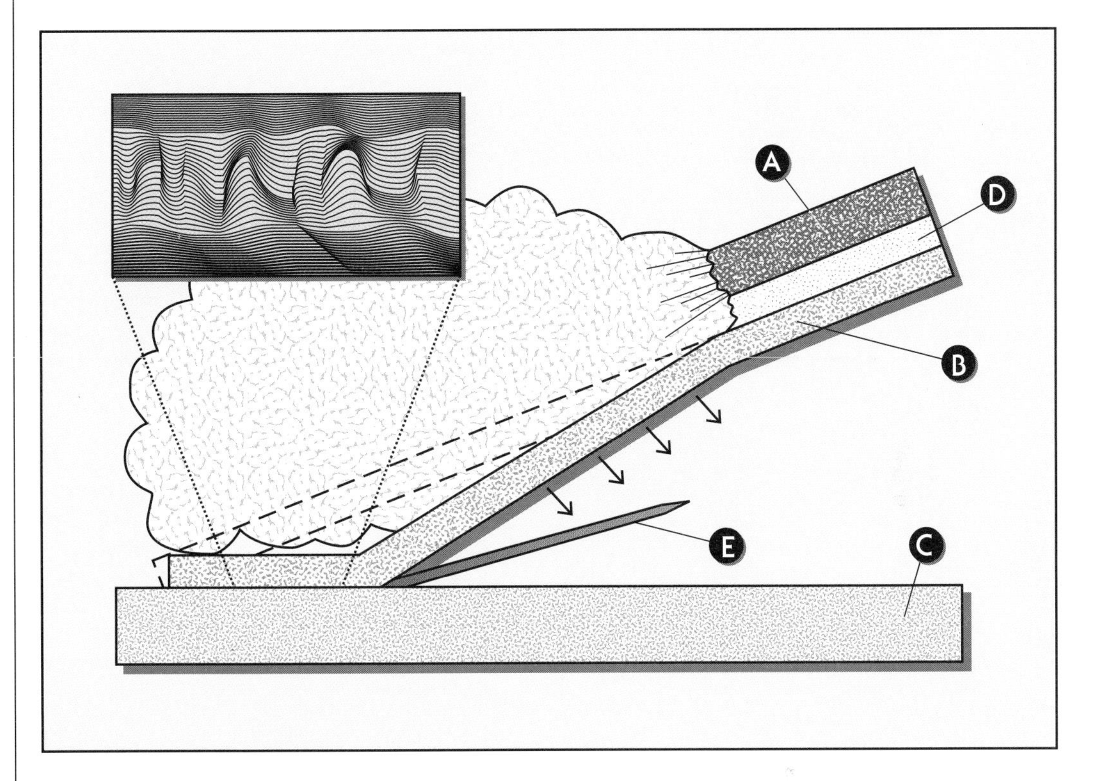

Many factories use explosives to join two sheets of metal more tightly than even the finest hand or machine welding job can join them. This technique, called explosive cladding, uses a layer of high explosive (A) to drive one sheet of metal (B) into the other (C). The top sheet of metal is shielded from the explosive charge by a layer of buffering material (D), which prevents the explosion from puncturing or otherwise damaging the two metal sheets. Forcing the two plates together this way causes their facing surfaces to briefly melt and swirl, locking them together when they cool.

Simply slamming the two sheets together, however, would not guarantee a tight-fitting weld. Tiny particles of dirt on either of the two plates, as well as air bubbles, could create pockets where rust could develop. In order to make the plate join as tightly as possible, the top sheet of metal and its explosive charge are set at an angle to the bottom sheet. This way, when the charge detonates, the melting surfaces of the two plates squeeze a small jet of molten metal (E) from between them. The molten jet carries away any surface contamination and pushes away all the air between the two plates.

One example of how explosive cladding is used is the U.S. quarter, which is made from a long-lasting "sandwich" of copper and nickel sheets. Explosive cladding also is used to join otherwise incompatible metals, such as steel and aluminum, which cannot be joined by ordinary welding.

a tighter bond than even the most skilled hand welder could create.

Some modern ships, especially warships, use aluminum in part of their construction to save weight, cut down on fuel consumption, and increase speed. But most of the rest of the ships are made out of steel, and aluminum and steel cannot normally be welded together because of their chemical makeup. This is where explosive welding comes in. Blast-welding a sheet of aluminum to a sheet of steel creates a composite that construction crews can weld to individual plates of each metal. By welding steel to steel and aluminum to aluminum, shipbuilders can make the various steel and aluminum sections of a ship hold together.

At the Energetic Materials Research and Testing Center near Socorro, scientists have used explosives to weld titanium and aluminum, two otherwise incompatible materials. The resulting alloy, titanium aluminide, is extremely strong and lightweight, and can withstand an amount of heat that would deform less resistant metals. Materials such as titanium aluminide might form the skins of future faster-than-sound airplanes, or even of the proposed space plane that would fly from America to Japan by way of low earth orbit.

Explosives can also be used to compress metals slightly, hardening them for jobs that create a great deal of wear. For example, Detasheet, a sheet of PETN-based explosive developed by Du Pont, is used to blast-toughen some types of railroad track, especially tracks that are used at switches. Just as blast-welding fuses two types of metal together, blast-hardening makes the metal in the tracks fuse and flow into itself, compressing the metal slightly.

Even metals that can be successfully welded can benefit from being welded by explosives. The finest job of conventional welding leaves tiny gaps between the two metals, opening the door to rust and other types of corrosion. Explosion welding, because it fuses the two metals together, eliminates any possibility of gaps. Some areas where explosive welding techniques already are being used include nuclear power plants, where explosive plugs fix tiny leaks in coolant and other pipes, and on some European high-speed rail lines, where tracks are welded together for smoother and faster rides.

At the Energetic Materials Research and Testing Center, a researcher lifts a cylinder containing explosively compacted titanium aluminide, a heat-resistant material that may one day form the skin of experimental space planes.

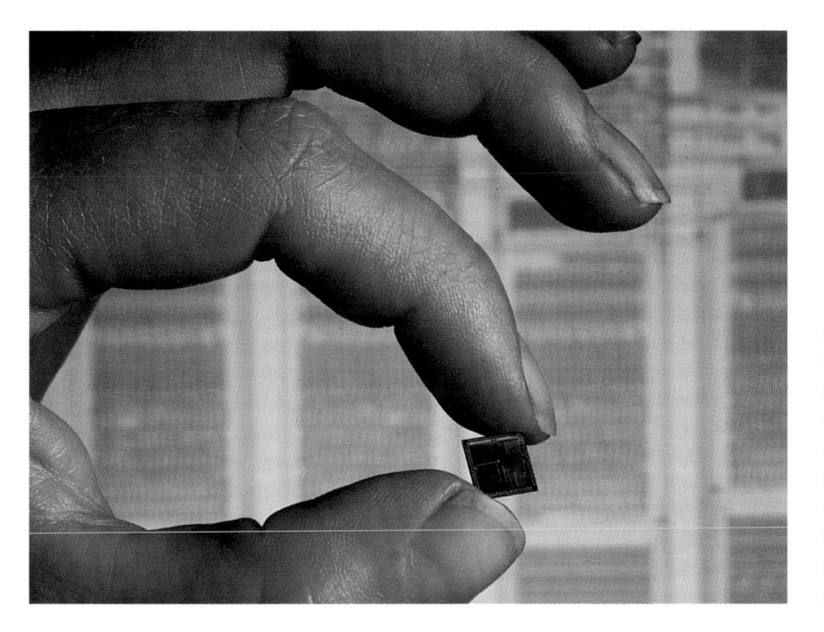

Silicon computer chips (pictured) can get very hot while running a high-powered computer. Scientists think that synthetic industrial diamonds created with the help of explosives might offer a good substitute for silicon.

Explosives and Computer Technology

Another beneficial use of explosives someday might be to create industrial diamonds that could be used to make computer chips. Computer chips, especially the latest high-powered central processors, can get very hot while they are operating. Silicon has been used to make these chips because it is a fairly cheap material that can quickly dissipate the heat the chips generate. But diamonds conduct heat far better than silicon, far better even than copper, silver, or aluminum, all of which are prized for their ability to warm and cool rapidly. A cheap way to produce diamonds for computer chips would mean that the chips could be made smaller and more powerful than those on the market today.

Experiments that use explosive charges to create synthetic industrial diamonds already are under way. The diamonds the explosives engineers are making do not look like the ones found in jewelry. Some of these diamonds look like small, black BB shot; others look like thin wafers of black glass or quartz. To make these experimental diamonds, the engineers start with laboratory-developed synthetic diamond powder made from graphite. They pack this powder into a small chamber, usually made of steel, and then surround it with plastic explosives. The shock and pressure of the detonation crush the chamber and force the powder into pellets of hard, black diamond. So far, these diamonds can conduct heat nearly as well as diamonds found in nature. Even so, the quality of the diamonds produced needs to be improved before they can be used as beds for computer circuitry. Impurities in the diamonds—such as small deposits of graphite that do not change into diamond—need to be eliminated to give the chips a long working life. Most researchers involved with creating these diamonds say it will take a few decades before the process will be ready for widespread use.

Explosives manufacturing methods also may play a role in developing future information superhighways, the proposed high-tech links between computers in the United States and around the world. At the research center in Socorro, engineers have used the pressure created by ANFO explosions to make superconducting ceramics out of ceramic powders. As the detonation wave travels through the powder, it squeezes the grains together hard enough for their surfaces to melt and mix together. The interiors of the grains, though, stay cool and retain their superconducting abilities. Many scientists and other experts believe that superconducting cables, together with fiber-optic lines, will carry signals between computers across the nation. The scientists involved with this type of research think that fusing powder with explosives will make setting up cable networks easier. All that work crews will have to do, some scientists think, is dig trenches for the cables, pour in the powder, and blast the powder into its final, ready-to-use form.

Same Explosives, Different Forms

Scientists are able to create industrial diamonds, superconducting ceramics, and advanced metal alloys in their laboratories because of the way they use explosives. In many experiments, scientists use shaped charges to focus the power of the explosives they use. A solid cone of slow-detonating high explosive is surrounded with a layer of faster-detonating explosive, forming an explosive lens. As the explosives detonate, most of the force is directed toward the bottom of the cone and whatever material is placed below it.

The detonation waves of the two layers of explosives arrive at the bottom of the lens at the same time. The shock wave, therefore, is almost perfectly flat, and affects all parts of the sample material at the same time and with the same force. This single, sudden burst of energy fuses the material, forming the desired ceramic, metal alloy, or other product the scientist wishes to produce.

Explosives manufacturing methods may play an important role in developing future information highways. Superconducting cables, created by the pressure of ANFO explosions, may someday carry signals between computers across the nation.

A planned explosion at the research center in Socorro sends sand and soil fifty feet into the air. Engineers at the center use the pressure created by such explosions to create synthetic industrial diamonds and superconducting ceramics.

Although the explosive lens technique has allowed scientists to create new materials, it also has allowed them to study how explosives work. This added understanding helps create new ideas about how future explosives might be used. Some of these possibilities include high-security checkbooks that fire ink charges through the checks if a code is not entered before the book is opened, and electronic time bomb pills for livestock that release medicine on a pre-timed pattern. There might even be a medical use for explosives in humans, such as a catheter tube that carries a tiny amount of explosive to break up gallstones. The ability of explosives to safely perform these tasks will depend on the knowledge gained in further explosives research. Knowing how an explosive or a combination of explosives goes from a stable state to total disruption, knowing how each event in the explosion takes place, will give humanity more control over the great forces contained in these little packages.

Glossary

ammonium nitrate: A compound of ammonia and nitrate used as an ingredient of some explosives and as a fertilizer.

black powder: A fast-burning (deflagrating) mixture of charcoal, sulfur, and saltpeter. Used mainly for artillery, firearms, and fireworks until the early 1900s. Also called gunpowder.

blasting: Using high explosives to break apart rock, demolish buildings, or to perform other tasks that require large amounts of force.

blasting agent: An explosive designed mainly for use in mines or on construction sites.

blasting cap: A small metal tube or container, filled with explosives, that is used to set off a larger charge of high explosives.

blast-welding: Using a high explosive charge to drive a sheet of one type of metal into a sheet of a different metal, fusing the two sheets. Used mainly for metals that cannot be welded using conventional techniques, such as aluminum and steel.

***brisance*:** The shattering effect of a detonation. Term used mainly to describe military explosives, which need to be able to penetrate the armor of enemy vehicles.

catalyst: In chemistry, a chemical that starts or speeds up a particular chemical reaction.

cellulose nitrate: A nitrated form of cotton or other plant materials, used mainly as a propellant for artillery and firearms.

charge: Any amount of explosive used in a construction blast or a military explosion.

deflagration: The process of burning something very fast.

detonation: An instantaneous combustion of the molecules in a material that produces a large shock wave and a large amount of gas.

detonation wave: A plane of force that is pushed through a mass of explosives during a detonation.

dynamite: A combination of nitroglycerin and kieselguhr, and later of ammonium nitrate, ethylene glycol dinitrate, and various fillers, used as a blasting agent.

explosive: Any material that can be made to generate a large amount of gas and a great deal of mechanical force in a brief time. There are three types of explosives: mechanical (such as boiling water in a closed container until it explodes), chemical (such as gunpowder, nitroglycerin, or dynamite), and nuclear.

fuse: A device used to set off an explosion.

gunpowder: See **black powder**.

high explosive: An explosive that detonates.

low explosive: An explosive that deflagrates.

molecular bond: The force that holds atoms together to form a molecule. Created by atoms in the molecule sharing electrons.

molecule: The smallest piece of a particular substance that can be identified as that substance. Formed by groups of atoms either of the same element (such as gold) or of different elements (such as salt, a combination of sodium and chlorine).

nitration: A process that combines nitrogen dioxide (NO_2) with an organic compound.

nitric acid: A strong fuming acid that is used to nitrate carbon-bearing organic compounds to make explosives.

nitrocellulose: A flammable compound made by nitrating cellulose, the material that makes up the walls of plant cells. Also called cellulose nitrate or guncotton.

nitroglycerin: An explosive liquid made by nitrating glycerol, which originally was a by-product of the soap industry.

plastic explosive: A light, stable, moldable type of extremely powerful explosive originally developed for U.S. Army demolition teams during World War II. Popular with terrorists because of its power, its light weight, and its relative invisibility to most forms of detection.

plastique: A type of plastic explosive.

primary explosive: A small charge of high explosive used to set off a larger amount of less sensitive, more powerful high explosive.

primer: Another term for primary explosive.

primer cord: A long, thin string of primary explosive, usually used in construction and building demolition work.

secondary explosive: A large charge of high explosive detonated using a primary explosive.

Semtex: A type of plastic explosive, consisting mostly of PETN and RDX, that was developed and made by Czechoslovakia. A very popular explosive among terrorists, mainly because it is so powerful.

sensitizer: A chemical that makes an explosive explode more easily by making it more sensitive to shock waves.

shaped charge: A charge of high explosives placed in a specific pattern in a weapon or in a civilian blasting device that is designed to focus the force of an explosion in a particular direction or at a particular point on a wall or a vehicle.

Sprengel explosive: An explosive made by mixing solid and liquid ingredients on the site where an explosion is to take place.

TNT: Trinitrotoluene. A powerful explosive used mostly by the military. Also was used in construction projects, such as dam building or tunneling, where an explosive that could shatter large amounts of rock was needed.

water gel: A type of explosive made by mixing ammonium nitrate, water, and various binders and sensitizers. Can be carried to blast sites and pumped into blast holes by tank trucks.

For Further Reading

Steven Ashley, "Dynamite Metals," *Popular Science,* March 1989.

Wayne Biddle, "It Must Be Simple and Effective," *Discover,* June 1986.

Wernher von Braun and Frederick I. Ordway III, *The Rockets' Red Glare.* Garden City, NY: Anchor Press/Doubleday, 1976.

Jerome Greer Chandler, "Plastic Terror," *The American Legion,* June 1986.

Donald Barr Chidsey, *Goodbye to Gunpowder.* New York: Crown Publishers, 1963.

Christopher Dickey, "Car Bombs: A Potent Weapon of Fear," *Newsweek,* March 8, 1993.

Donald Dale Jackson, "While He Expected the Worst, Nobel Hoped for the Best," *Smithsonian,* November 1988.

George Plimpton, *Fireworks: A History and Celebration.* Garden City, NY: Doubleday, 1984.

Donovan Webster, "'Out There Is a Bomb with Your Name,'" *Smithsonian,* February 1994.

Richard Wolkomir, "Boomtown," *Discover,* August 1989.

Works Consulted

Kenneth W. Banta, "The Arms Merchants' Dilemma," *Time*, April 2, 1990.

Erik Bergengren, *Alfred Nobel*. London: Thomas Nelson and Sons, 1962.

Blaster's Handbook: A Manual Describing Explosives and Practical Methods of Use. Wilmington, DE: E. I. du Pont de Nemours, 1967 and 1980.

"Bomb-Eating Bugs" ("Breakthroughs" column), *Discover*, August 1993.

Malcolm W. Browne, "Making Explosives Safe Is Aim of Testing Center," *The New York Times*, September 21, 1993 (late edition).

Melvin A. Cook, *The Science of High Explosives*. Huntington, NY: Robert E. Krieger Publishing Company, 1971.

Rae Corelli, "A Libyan Connection," *Maclean's*, November 25, 1991.

Joe Dannenberg, *Contemporary History of Industrial Explosives in America*. Westport, WA: ABA Publishing Company, 1978.

William C. Davis, "The Detonation of Explosives," *Scientific American*, May 1987.

Bryan Di Salvatore, "Vehement Fire-I" ("A Reporter At Large"), *The New Yorker*, April 27, 1987.

———"Vehement Fire-II" ("A Reporter At Large"), *The New Yorker*, May 4, 1987.

"The Electronic Look of Explosives" (Chemistry column), *Science News*, August 22, 1987.

Michael Evlanoff and Marjorie Flour, *Alfred Nobel: The Loneliest Millionaire.* Anderson, Ritchie & Simon, 1969.

"Explosives Technology Opens Range of Design Applications," *Design News,* May 2, 1988.

Arthur Fisher, "Bang! You're a Diamond" ("Science Newsfront" column), *Popular Science,* January 1989.

———, "Exploding Faster" ("Science Newsfront" column), *Popular Science,* August 1987.

Joseph Gies, *Adventure Underground: The Story of the World's Great Tunnels.* Garden City, NY: Doubleday, 1962.

Tony Gray, *Champions of Peace.* Paddington Press, 1976.

"International Controls to Regulate Plastic Explosives Adopted in Montreal," *UN Chronicle,* June 1991.

Mary Janigan, "A Tragedy's Haunting Legacy," *Maclean's,* June 23, 1986.

D. Keith Mano, "The Making of a Terrorist," *National Review,* June 19, 1987.

"Method Uses Bugs to Eat TNT" ("R&D News" column), *R&D Magazine,* February 1993.

Michael Newton, *Armed and Dangerous.* Cincinnati, OH: Writer's Digest Books, 1990.

Thomas Parrish, *The Simon and Schuster Encyclopedia of World War II.* New York: Simon and Schuster, 1978.

Jonathan Rose, "What 19th-Century Terror Tells Us About Today's," *Scholastic Update*, May 16, 1986.

Gösta E. Sandström, *Tunnels*. New York: Holt, Rinehart and Winston, 1962.

Gary Slutsker, "Kaboom!" ("Science & Technology" column), *Forbes*, February 20, 1989.

Marcelle M. Soviero, "Bacteria That Eat TNT," *Popular Science*, November 1989.

Tim Stevens, "Applications Explode (Literally) in Energetic Materials," *Industry Week*, March 15, 1993.

M. Mitchell Waldrop, "FAA Fights Back on Plastic Explosives," *Science*, January 13, 1989.

Douglas Waller, "Caught by Surprise," *Newsweek*, March 8, 1993.

"Wall Street Tragedy" ("Talk of the Town" column), *The New Yorker*, March 15, 1993.

Index

About the Author

Sean M. Grady has had a varied career both as a journalist and as a free-lance writer. While still in college, he wrote for the *Los Angeles Times* and worked as a general assignment reporter for the City News Service of Los Angeles, a regional news wire. In the late 1980s, he served as business editor of the *Olympian*, a Gannett newspaper in Olympia, Washington. His writing credits with Lucent Books include *The Importance of Marie Curie*, *Submarines*, *Illiteracy*, and *Plate Tectonics*. Grady lives near Reno, Nevada.

Picture Credits

Cover photo by Jim Zipp/Photo Researchers, Inc.
AP/Wide World Photos, 55, 62, 74
Charlie Archambault, *U.S. News & World Report*, 53
The Bettmann Archive, 15, 20, 21, 71
Bureau of Reclamation, 35 (bottom)
© Charles Feil 1984/UNIPHOTO Picture Agency, 77
Jean-Marc Giboux/© 1989 *Discover Magazine*, 79
© 1992 Scott Goldsmith, 73 (bottom)
Henry Grokinsky/© 1986 *Discover Magazine*, 51
Hagley Museum and Library, 16, 22, 27, 33, 34, 37 (both), 58 (both), 61 (both), 64, 65 (both)
Leo de Wys Inc./Paul Kennedy, 63
Leo de Wys Inc./Randy Taylor, 69 (all)
Library of Congress, 14, 23, 44
Los Alamos National Laboratory, 54
NASA, 68 (top)
National Archives, 45, 47, 48, 50 (bottom), 52
Ray Nelson/© 1989 *Discover Magazine*, 76
North Wind Picture Archives, 17 (top), 18, 24, 28, 30, 38, 39, 43
Reuters/Bettmann, 56, 68 (bottom)
Sandia National Laboratories, 72 (bottom), 73 (top)
Stock Montage, Inc., 13 (both), 17 (bottom), 19, 25, 29, 32, 40, 49
UNIPHOTO Picture Agency, 35 (top), 67, 72 (top), 78
UPI/Bettmann, 50 (top), 60